ARITHMETIC

Richard L. Steinhoff
Department of Mathematics
Modesto Junior College

McGraw-Hill Book Company
New York • St. Louis • San Francisco • Auckland • Bogotá • Düsseldorf
Johannesburg • London • Madrid • Mexico • Montreal • New Delhi
Panama • Paris • São Paulo • Singapore • Sydney • Tokyo • Toronto

Copyright © 1978 by McGraw-Hill, Inc. All rights reserved.
Printed in the United States of America. No part of this publication
may be reproduced, stored in a retrieval system, or transmitted, in any
form or by any means, electronic, mechanical, photocopying, recording, or
otherwise, without the prior written permission of the publisher.

5 6 7 8 9 0 EBEB 88 87 86 85 84 83

Library of Congress Cataloging in Publication Data

Steinhoff, Richard L
   Arithmetic.

   Includes index.
   1. Arithmetic—1961–   I. Title.
QA107.S73   513   77-6417
ISBN 0-07-061127-0

This book was set in Times Roman by Monotype Composition Company, Inc.
The editors were A. Anthony Arthur and Michael Gardner;
the designers were J. Paul Kirouac and
Craigwood Phillips, A Good Thing, Inc.;
the production supervisor was Dennis J. Conroy.
The drawings were done by Monotype Composition Company, Inc.

CONTENTS

	Preface	ix

Unit 1 **Whole-Number Arithmetic**

Section 1-1	Place Value and Expanded Form	3
Section 1-2	Addition of Whole Numbers	9
Section 1-3	Subtraction of Whole Numbers	15
Section 1-4	Multiplication of Whole Numbers	21
Section 1-5	Division of Whole Numbers	27
	Unit 1 Test	35

Unit 2 **Prime Factorization, Equivalent Fractions, Least Common Multiple**

Section 2-1	Prime Factorization	39
Section 2-2	Reducing Fractions	49
Section 2-3	Equivalent Fractions with Larger Denominator	57
Section 2-4	Least Common Multiples	61
	Unit 2 Test	67

Unit 3 **Arithmetic of Common Fractions**

Section 3-1	Multiplication of Fractions	71
Section 3-2	Division of Fractions	81
Section 3-3	Addition of Fractions	89
Section 3-4	Subtraction of Fractions	103
	Unit 3 Test	111

Unit 4 **Mixed-Number Arithmetic**

Section 4-1	Expressing Mixed Numbers as Improper Fractions	115
Section 4-2	Expressing Improper Fractions as Mixed Numbers	119
Section 4-3	Multiplication and Division of Mixed Numbers	123

Section 4-4	Addition of Mixed Numbers	129
Section 4-5	Subtraction of Mixed Numbers	137
	Applications	143
	Multiplication and Division	143
	Addition and Subtraction	146
	Unit 4 Test	151

Unit 5 Decimal Arithmetic

Section 5-1	Decimal Numbers	155
Section 5-2	Addition and Subtraction of Decimal Numbers	161
Section 5-3	Multiplication of Numbers with Decimal Part	167
Section 5-4	Rounding Decimal Numbers	171
Section 5-5	Division of Decimal Numbers	177
Section 5-6	Writing Fractions as Equivalent Decimals	185
	Applications	191
	Unit 5 Test	197

Unit 6 Simple Equations

Section 6-1	Equations of the Form $ax = c$ and $ax/b = c$	201
Section 6-2	Equations of the Form $x - a = b$ and $x + a = b$	209
	Unit 6 Test	213

Unit 7 Percents

Section 7-1	Percent as a Number	217
Section 7-2	Basic Problems with Percents	225
Section 7-3	Application of Percent	235
	Unit 7 Test	245

Unit 8 Arithmetic of Signed Numbers

Section 8-1	Negative Numbers and Inequalities	249

Section 8-2	Addition and Subtraction of Two Signed Numbers	255
Section 8-3	Addition and Subtraction Problems Involving Several Numbers	261
Section 8-4	Multiplication and Division of Signed Numbers	267
	Unit 8 Test	277

Unit 9 Order of Operations

Section 9-1	Problems Involving Multiplication, Division, Addition, and Subtraction	281
Section 9-2	Parentheses and Brackets Used to Group Operations	289
Section 9-3	Division Line	297
	Unit 9 Test	305

Unit 10 Formulas

Section 10-1	Basic Formulas	311
Section 10-2	Problems Involving More Than One Formula	323
Section 10-3	Symbolic Representation	333
Section 10-4	Mathematical Statements	341
Section 10-5	Problems Involving More Than One Verbal Formula	349
	Unit 10 Test	359

Unit 11 Denominate Numbers

Section 11-1	Conversion Factors	365
Section 11-2	The Metric System	377
Section 11-3	The English System	393
Section 11-4	English to Metric and Metric to English	405
	Table of Equivalences	408
Section 11-5	Applications: Speed, Density, Concentration, and Cost per Unit	417
	Unit 11 Test (Sections 1 through 4)	433
	Unit 11 Test (Section 5)	437

Unit 12 **Exponents and Roots**

Section 12-1	Positive Exponents	443
Section 12-2	Negative Exponents	451
Section 12-3	Scientific Notation	459
Section 12-4	Square and Cube Roots of Whole Numbers	473
Section 12-5	Square and Cube Roots of Fractions	481
	Unit 12 Test	491
	Answers to Odd-numbered Exercises	493
	Index	517

PREFACE

Arithmetic was written for community-college students with the intent of covering a wide range of needs. Several of the earlier units provide review and, depending on the needs of the student, may be omitted. The main emphasis of the text is on how to work problems and on applications. Multistep problems tend to be difficult for most students. For this reason, the units on formulas and order of operations stress the organization needed to set up and solve multistep problems. Word problems giving applications in many areas are placed throughout the text, starting in Unit 4. The extent to which precise definitions and demonstrations are given will provide the student with a solid background for a beginning algebra course.

Each section consists of four or five parts:

1. A brief description of concepts, together with definitions and rules, at the beginning of each section. Mathematical demonstrations are included here only when their inclusion is essential to the student's basic understanding of what is going on.
2. Sample problems.
3. A sample set in which students complete solutions to partially worked problems.
4. In many sections the sample set is followed by supplementary information intended to add clarity and interest, or by mathematical justification of rules.
5. Exercises.

The first part of each section is intended to provide a direct route to the solution of problems. It is for this reason that the more complicated justification of rules is placed after the sample set. Laws of mathematics are used only to the extent of showing how laws form the foundation of arithmetic. Students are not asked to identify laws used in a demonstration.

I am grateful for the help given by reviewers Darrell Top, and especially Ken Skeen. I am also grateful to my wife, Elaine, who typed (again and again) the manuscript.

Richard L. Steinhoff

Section 1-1 **Place Value and Expanded Form**

It is a credit to the ingenuity of mankind that every number can be written using the 10 **digits** 0, 1, 2, 3, 4, 5, 6, 7, 8, and 9. For example, the number 3,475 is represented using the four digits 3, 4, 7, and 5.

In the number 3,475, the digit 3 is in the thousands' place, 4 is in the hundreds' place, 7 is in the tens' place, and 5 is in the ones' place. This indicates that we have 3 thousands, 4 hundreds, 7 tens, and 5 ones. The number 3,475 can therefore be written in the **expanded form**

$$3,475 = 3 \text{ thousands} + 4 \text{ hundreds} + 7 \text{ tens} + 5 \text{ ones}$$
or $$3,475 = 3,000 + 400 + 70 + 5$$

In order to write a number in expanded form, it is important to know the **place value** associated with each digit in the number. The place value of each digit in the number 3,745,829 is indicated below.

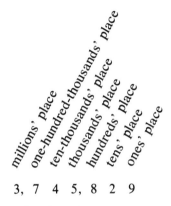

In expanded form, the number is written

$$3,745,829 = 3,000,000 + 700,000 + 40,000 + 5,000 + 800 + 20 + 9$$

Sample Problem 1-1

Write 62,721 in expanded form.

Solution:

$$62,721 = 60,000 + 2,000 + 700 + 20 + 1$$

Sample Problem 1-2

Write 3,705 in expanded form.

Solution: The digit 0 in the tens' place tells us we have no tens. This is indicated in expanded form by writing only the number of thousands, hundreds, and ones.

$$3,705 = 3,000 + 700 + 5$$

Exponential Form Whole numbers that form a product are called **factors** of that product. For example, in the product

$$3 \times 4 = 12$$

the numbers 3 and 4 are factors. A product in which the same factor is repeated may be written in **exponential form**.

Example $\qquad 3 \times 3 \times 3 \times 3 = 3^4 \qquad$ **in exponential form**

The exponential form 3^4 is called the fourth **power** of 3. The number 3 is the **base** of the power, and the number 4 is called the **exponent**. Note that the exponent counts the number of times the base appears as a factor in the product.

Several powers of 10 are written below:

$$10 = 10^1 \qquad \text{10 to the first power}$$
$$10 \times 10 = 100 = 10^2 \qquad \text{10 to the second power}$$
$$10 \times 10 \times 10 = 1{,}000 = 10^3 \qquad \text{10 to the third power}$$
$$10 \times 10 \times 10 \times 10 = 10{,}000 = 10^4 \qquad \text{10 to the fourth power}$$

The number 3,245 is written below in expanded form using powers with base 10:

$$3{,}245 = 3{,}000 + 200 + 40 + 5$$
$$= (3 \times 1{,}000) + (2 \times 100) + (4 \times 10) + 5$$
$$= (3 \times 10^3) + (2 \times 10^2) + (4 \times 10^1) + 5$$

Sample Problem 1-3

Solution:

Write 3,097 in expanded form using powers of 10.

$$3{,}097 = (3 \times 1{,}000) + (9 \times 10) + 7$$
$$= (3 \times 10^3) + (9 \times 10^1) + 7$$

Sample Problem 1-4

Solution:

Write 87,563 in expanded form using powers of 10.

$$87{,}563 = (8 \times 10{,}000) + (7 \times 1{,}000) + (5 \times 100) + (6 \times 10) + 3$$
$$= (8 \times 10^4) + (7 \times 10^3) + (5 \times 10^2) + (6 \times 10^1) + 3$$

In the problems below, write a digit in each box to make each problem complete. Completed problems appear on page 6.

Sample Set

(a) $32 = \boxed{}0 + 2$

$\qquad = 3 \times 10^{\boxed{}} + 2$

(b) $529 = 5\boxed{}\boxed{} + 2\boxed{} + \boxed{}$

$\qquad = 5 \times 10^{\boxed{}} + 2 \times 10^{\boxed{}} + \boxed{}$

(c) $7{,}436 = \boxed{}000 + \boxed{}00 + 3\boxed{} + 6$

$\qquad = 7 \times 10^{\boxed{}} + 4 \times 10^{\boxed{}} + 3 \times 10^{\boxed{}} + 6$

Sec. 1-1 Place Value and Expanded Form

(d) $8,030 = \boxed{}000 + 3\boxed{}$

$ = 8 \times 10^{\boxed{}} + 3 \times 10^{\boxed{}}$

(e) $796,295 = 7\boxed{}\boxed{},\boxed{}\boxed{}\boxed{} + \boxed{}$

$ + 6,000 + \boxed{}00 + \boxed{}0 + \boxed{}$

$ = 7 \times 10^{\boxed{}} + 9 \times 10^{\boxed{}} + 6 \times 10^{\boxed{}} + 2 \times 10^{\boxed{}}$

$ + 9 \times 10^{\boxed{}} + \boxed{}$

Background Information

It was stated at the beginning of this section that it is a credit to the ingenuity of mankind that every number can be represented using the 10 digits 0, 1, 2, 3, 4, 5, 6, 7, 8, and 9. One can only imagine the difficulties we would encounter if we attempted to use a different symbol to represent each number. Since there is no largest number, this would be impossible.

It might be interesting to guess how our number system had its beginning. Possibly, at the end of a day, a shepherd was counting his sheep. He might have started counting sheep on his fingers. Each time he got to 10, and ran out of fingers, he might have set a stone down. Each stone would then represent 10 sheep. If the shepherd had a large herd of sheep, it is possible that after setting down 10 stones, the shepherd set down a larger stone representing 100 sheep.*

The important fact is that we group 10 units together to make one ten, 10 tens together to make one hundred, 10 hundreds together to make one thousand, and so on. It is this grouping into powers with base 10 that makes our number system a **base-10** system.

From the time of our shepherd, many centuries undoubtedly passed before someone invented symbols to represent the 10 digits and actually wrote a number. Although setting digits next to each other in places (ones' place, tens' place, hundreds' place, and so on) is a simple idea for us to understand, we must consider it an outstanding achievement for the person who first thought of doing this.

Writing a number such as 328 in expanded form,

$$328 = 3 \text{ hundreds} + 2 \text{ tens} + 8 \text{ ones}$$

demonstrates our grouping of units, tens, and hundreds. The expanded form of a number will be used in future sections of this unit to help explain many of the familiar operations we use when adding, subtracting, multiplying, and dividing numbers.

*Eves, Howard, "An Introduction to the History of Mathematics," Holt, Rinehart and Winston, Inc., New York, 1972, pp. 7–8.

Completed Problems

(a) $32 = \boxed{3}0 + 2$
$= 3 \times 10^{\boxed{1}} + 2$

(b) $529 = 5\boxed{0}\boxed{0} + 2\boxed{0} + \boxed{9}$
$= 5 \times 10^{\boxed{2}} + 2 \times 10^{\boxed{1}} + \boxed{9}$

(c) $7,436 = \boxed{7}000 + \boxed{4}00 + 3\boxed{0} + 6$
$= 7 \times 10^{\boxed{3}} + 4 \times 10^{\boxed{2}} + 3 \times 10^{\boxed{1}} + 6$

(d) $8,030 = \boxed{8}000 + 3\boxed{0}$
$= 8 \times 10^{\boxed{3}} + 3 \times 10^{\boxed{1}}$

(e) $796,295 = 7\boxed{0}\boxed{0},\boxed{0}\boxed{0}\boxed{0} + \boxed{90,000}$
$+ 6,000 + \boxed{2}00 + \boxed{9}0 + \boxed{5}$
$= 7 \times 10^{\boxed{5}} + 9 \times 10^{\boxed{4}} + 6 \times 10^{\boxed{3}} + 2 \times 10^{\boxed{2}}$
$+ 9 \times 10^{\boxed{1}} + \boxed{5}$

Name: _____

Class: _____

Exercises for Sec. 1-1 Write each number in expanded form, as shown in sample problems 3 and 4.

1. 29 =

2. 53 =

3. 734 =

4. 804 =

5. 930 =

6. 2,572 =

7. 6,081 =

8. 1,700 =

9. 5,697 =

10. 31,632 =

11. 55,091 =

12. 70,805 =

13. 92,007 =

14. 80,300 =

15. 375,216 =

16. 832,504 =

17. 100,253 =

18. 705,600 =

19. 8,752,546 =

20. 2,340,907 =

Section 1-2 Addition of Whole Numbers

The process used to add whole numbers will be reviewed in this section. This process will be justified, using the commutative and associative laws, at the end of the section.

An addition problem may be written horizontally (on the line) or vertically (up and down).

Horizontal: $37 + 28 + 19$

Vertical:
$$\begin{array}{r} 37 \\ 28 \\ +19 \end{array}$$

Below you will find the problem worked:

$$\begin{array}{r} 2 \leftarrow \text{carry 2} \\ 37 \\ 28 \\ +19 \\ \hline 84 \end{array} \qquad \begin{array}{l} 7 + 8 + 9 = 24 \\ = 2 \text{ tens} + 4 \text{ ones,} \\ \text{so carry 2 tens} \end{array}$$

The process of carrying a number will be discussed further at the end of this section.

In an addition problem, the numbers being added are called **addends.** The answer obtained is called the **sum.**

$$37 + 28 + 19 = 84$$
     addends   sums

Sample Problem 1-5 Add $389 + 476 + 829$.

Solution:

Step 1. Write the problem vertically.

Step 2. Add the digits in the ones' column. Since $9 + 6 + 9 = 24$, write 4 below the ones' column and carry 2:

$$\begin{array}{r} 2 \\ 389 \\ 476 \\ 829 \\ \hline 4 \end{array}$$

Step 3. Add the digits in the tens' column. Since $2 + 8 + 7 + 2 = 19$, write 9 below the tens' column and carry 1:

$$\begin{array}{r} 12 \\ 389 \\ 476 \\ 829 \\ \hline 94 \end{array}$$

Step 4. Add the digits in the hundreds' column. Since $1 + 3 + 4 + 8 = 16$, write 6 below the hundreds' column and place a 1 in the thousands' place of the sum:

$$\begin{array}{r} 12 \\ 389 \\ 476 \\ 829 \\ \hline 1{,}694 \end{array}$$

10 Whole-Number Arithmetic

Complete the problems in the sample set. Completed problems are given on page 12.

Sample Set

(a)
```
    2 4
  + 3 2
  -----
    5 ☐
```

(b)
```
      ☐
    7 3
  + 1 9
  -----
    9 2
```

(c)
```
      ☐
    6 4
  + 2 8
  -----
    9 ☐
```

(d)
```
      ☐
    5 8
  + 7 9
  -----
  ☐ 3 7
```

(e)
```
      ☐
    7 8
  + 8 9
  -----
  1 ☐ ☐
```

(f)
```
    ☐ ☐
    3 7 0
      8 6
  + 4 6 9
  -------
    ☐ 2 5
```

(g)
```
      ☐ ☐
      8 4 7
      2 8 9
  +   5 7 1
  ---------
    1, ☐ 0 ☐
```

(h)
```
        ☐
        7 7
        8 6
        2 9
  +     3 8
  ---------
      ☐ 3 ☐
```

(i)
```
      ☐ ☐ ☐
      3, 4 8 9
      2, 3 9 3
  +   8, 9 6 5
  -----------
    ☐ 4, 8 ☐ 7
```

The Associative and Commutative Laws of Addition

Parentheses are used in mathematics to indicate the order in which operations are to be performed. **It is accepted that the operations inside parentheses are to be performed first.**

Observe that the problems below have the same answer:

$$3 + (2 + 4) = 3 + 6 = 9$$
$$(3 + 2) + 4 = 5 + 4 = 9$$

This result can be indicated by writing the single equality

$$3 + (2 + 4) = (3 + 2) + 4$$

Using letters a, b, and c to represent numbers, this observation can be stated as a general law.

Associative Law of Addition

For any numbers a, b, and c,
$a + (b + c) = (a + b) + c$

Have you ever used the associative law? The answer is yes! As an example, the law is used every time you carry a number in an addition problem. This is demonstrated below:

$$
\begin{aligned}
37 &= 3 \text{ tens} + 7 \text{ ones} \\
28 &= 2 \text{ tens} + 8 \text{ ones} \\
+19 &= 1 \text{ ten } + 9 \text{ ones} \\
\hline
& 6 \text{ tens} + 24 \text{ ones} = 6 \text{ tens} + (2 \text{ tens} + 4 \text{ ones}) \\
& \phantom{6 \text{ tens} + 24 \text{ ones}} = (6 \text{ tens} + 2 \text{ tens}) + 4 \text{ ones} \quad \rightarrow \textbf{associative law} \\
& \phantom{6 \text{ tens} + 24 \text{ ones}} = 8 \text{ tens} + 4 \text{ ones} \\
& \phantom{6 \text{ tens} + 24 \text{ ones}} = 84
\end{aligned}
$$

The 2 tens were originally associated with the 4 ones as part of 24. The associative law justified regrouping the 2 tens with the 6 tens. This regrouping is actually the process of carrying a 2 to the tens' column. The 2 we carry is really 2 tens.

The commutative law of addition states numbers added in either order have the same sum. For example,

$$3 + 2 = 5 \quad \text{and} \quad 2 + 3 = 5$$

In a general form, this law is stated below.

Commutative Law of Addition

For any numbers a and b,
$$a + b = b + a$$

The commutative and associative laws are used when numbers in an addition problem are rearranged and added in a different order. This fact will be used later.

12 Whole-Number Arithmetic

Completed Problems

(a)
```
   2 4
 + 3 2
 -----
   5 [6]
```

(b)
```
   [1]
   7 3
 + 1 9
 -----
   9 2
```

(c)
```
   [1]
   6 4
 + 2 8
 -----
   9 [2]
```

(d)
```
     [1]
     5 8
   + 7 9
   -----
   [1] 3 7
```

(e)
```
     [1]
     7 8
   + 8 9
   -----
   1 [6][7]
```

(f)
```
   [2][1]
   3 7 0
     8 6
 + 4 6 9
 -------
   [9] 2 5
```

(g)
```
     [2][1]
     8 4 7
     2 8 9
   + 5 7 1
   -------
   1,[7] 0 [7]
```

(h)
```
     [3]
     7 7
     8 6
     2 9
   + 3 8
   -----
   [2] 3 [0]
```

(i)
```
      [1][2][1]
      3, 4 8 9
      2, 3 9 3
    + 8, 9 6 5
    ---------
    [1] 4, 8 [4] 7
```

Name: _____

Class: _____

Exercises for Sec. 1-2 Compute the following sums:

1. 36
  + 23

2. 48
  + 27

3. 84
  + 63

4. 57
  + 29

5. 67
  + 98

6. 71
  + 89

7. 232
  + 346

8. 258
  + 306

9. 583
  + 255

10. 743
  + 626

11. 853
  + 798

12. 38
  22
  + 37

13. 53
  72
  + 64

14. 25
  93
  + 74

15. 72
  50
  + 38

16. 48
  97
  + 59

17. 86
  80
  + 52

18. 102
  321
  + 435

19. 727
  682
  + 443

20. 869
  236
  + 541

21. 935
  782
  + 414

22. 769
  203
  + 97

23. 7
  9
  2
  8
  + 6

24. 12
  32
  45
  79
  + 63

25. 92
     71
     36
     28
    +17

26. 75
     20
     37
     64
    +19

27. 36
      7
     28
     53
    +21

28. 915
      82
     549
     360
    + 76

29. 8
     7
     1
     3
     7
    +6

30. 7
     9
     3
     2
     5
    +8

31. 73
     56
     27
     19
     30
    +82

32. 85
     15
      8
     26
     73
    + 4

33. 821
      34
     805
     127
      63
    + 92

34. 7,235
     8,567
     3,415
     7,062
    13,847
    + 5,876

35. 5
     2
     9
     6
     7
     8
    +2

36. 21
     32
     75
     63
     94
     28
    +37

37. 73
    205
    392
    567
     28
    347
    + 55

38. 81
      9
     32
     73
     54
      9
      7
    +16

39. 52
     36
     54
     29
     30
     81
     17
    +65

40. 83
     97
    143
     28
    396
    250
    537
    +418

Section 1-3 Subtraction of Whole Numbers

Subtraction is the **inverse operation** of addition. The word *inverse* suggests the notion of opposite. In addition and subtraction, the ideas of "adding to" and "taking away" can be thought of as opposite procedures or inverse operations. There is another way to view an inverse operation. In the problems

$$3 + 4 = 7$$
and
$$7 - 4 = 3$$

the subtraction operation in the second problem can be thought of as "undoing" the addition operation in the first. Starting with 3 in the first problem, an addition operation was performed to arrive at 7. In the second, a subtraction operation was used in order to get back to 3. In this sense the operations are opposite, or inverse.

In a subtraction problem, each number has a special name. These names are indicated below:

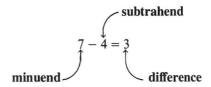

The fact that addition and subtraction are inverse operations provides a method for checking subtraction. For the subtraction to be correct, the sum of the difference and subtrahend must equal the minuend.

Example $\qquad 7 - 4 = 3 \qquad$ minuend − subtrahend = difference
Check: $\quad 3 + 4 = 7 \qquad$ difference + subtrahend = minuend

Below you will find a subtraction problem worked out.

$$\begin{array}{r} 896 \\ -324 \\ \hline 572 \end{array}$$

Note that the digits in each column have been subtracted.

The next example shows a case in which a number must be borrowed.

$$\begin{array}{r} {\scriptstyle 8\ 13} \\ \cancel{9}\cancel{3} \\ -67 \\ \hline 26 \end{array}$$

Here 7 is greater than 3, so 1 (really 10) was borrowed from the 9 in the tens' column. One uses the associative law of addition when borrowing a number.

$$93 = 9 \text{ tens} + 3 \text{ ones}$$
$$-67 = -(6 \text{ tens} + 7 \text{ ones})$$

$$= \quad (8 \text{ tens} + 1 \text{ ten}) + 3 \text{ ones}$$
$$= -\underline{(6 \text{ tens} \qquad\quad + 7 \text{ ones})}$$

← note the 1 ten grouped with the 8 tens

$$= \quad 8 \text{ tens} + (1 \text{ ten} + 3 \text{ ones})$$
$$= -\underline{(6 \text{ tens} + \qquad\quad 7 \text{ ones})}$$

the associative law is used when regrouping the 1 ten with the 3 ones

$$= \quad 8 \text{ tens} + 13 \text{ ones}$$
$$= -\underline{(6 \text{ tens} + \;\; 7 \text{ ones})}$$
$$2 \text{ tens} + 6 \text{ ones} \;\; = 26$$

this regrouping is the process of borrowing a number

Sample Problem 1-6

Solution:

Subtract 837 − 462.

Step 1. Write the problem vertically:

$$\begin{array}{r} 837 \\ -462 \\ \hline \end{array}$$

Step 2. Subtract the numbers in the ones' column:
7 − 2 = 5   write 5
(Since 2 is less than 7, we do not borrow.)

$$\begin{array}{r} 837 \\ -462 \\ \hline 5 \end{array}$$

Step 3. In the tens' column, 6 is greater than 3; so borrow 1 from the hundreds' column:
13 − 6 = 7   write 7

$$\begin{array}{r} {}^{7\;13}\\ 8\;\cancel{3}\;7 \\ -4\;6\;2 \\ \hline 7\;5 \end{array}$$

Step 4. Since 1 was borrowed from 8 to leave 7, find:
7 − 4 = 3   write 3

$$\begin{array}{r} {}^{7\;13}\\ \cancel{8}\;\cancel{3}\;7 \\ -4\;6\;2 \\ \hline 3\;7\;5 \end{array}$$

Check: 375 + 462 = 837.

Complete the problems in the sample set. Completed problems appear on page 18.

Sample Set

(a)
```
    4 9
  - 4 6
  -----
    ☐
```

(b)
```
    8 5
  - 2 4
  -----
   ☐☐
```

(c)
```
   ☐☐
    8̸ 4̸
  - 7 9
  -----
      ☐
```

(d)
```
   ☐☐
    5̸ 2
  - 2 6
  -----
    ☐ 6
```

(e)
```
   ☐☐
    7̸ 3̸ 8
  - 2 8 3
  -------
    ☐ 5 5
```

(f)
```
   ☐☐
    6̸ 2̸ 9
  - 2 6 3
  -------
    3 ☐☐
```

(g)
```
         ☐☐
    3, 6 8̸ 2̸
  -    2 3 5
  ----------
    3, ☐ 4 7
```

(h)
```
     ☐☐☐
     5̸, 3̸ 2̸ 8
  - 2,    8 7 0
  ------------
     ☐ 4 5 8
```

(i)
```
     ☐☐☐
     3̸, 2̸ 5̸ 7
  - 1,    4 5 2
  ------------
     1, ☐☐☐
```

18 Whole-Number Arithmetic

Completed Problems

(a)
```
   4 9
-  4 6
-----
     [3]
```

(b)
```
   8 5
-  2 4
-----
   [6][1]
```

(c)
```
   [7][14]
   8̸ 4̸
-  7 9
-----
       [5]
```

(d)
```
   [4][12]
   5̸ 2̸
-  2 6
-----
   [2] 6
```

(e)
```
   [6][13]
   7̸ 3̸ 8
-  2 8 3
-------
   [4] 5 5
```

(f)
```
   [5][12]
   6̸ 2̸ 9
-  2 6 3
-------
   3 [6][6]
```

(g)
```
         [7][12]
   3, 6 8̸ 2̸
-     2 3 5
-----------
   3, [4] 4 7
```

(h)
```
      [4][12][12]
   5̸, 3̸ 2̸ 8
-  2, 8 7 0
-----------
   [2] 4 5 8
```

(i)
```
      [2][11][11]
   3̸, 2̸ 8̸ 7
-  1, 4 5 2
-----------
   1, [7][6][5]
```

Name: _____

Class: _____

Exercises for Sec. 1-3 Compute the differences.

1.   18
   − 7

2.   28
   − 14

3.   36
   − 18

4.   80
   − 38

5.   94
   − 69

6.   78
   − 38

7.   46
   − 43

8.   72
   − 30

9.   60
   − 47

10.   51
    − 49

11.   73
    − 26

12.   84
    − 79

13.   400
    − 76

14.   597
    − 263

15.   629
    − 337

16.   276
    − 179

17.   750
    − 328

18.   563
    − 487

19. 892
    − 884

20. 8,436
    − 5,215

21. 9,321
    − 3,808

22. 7,839
    − 4,296

23. 5,362
    − 3,563

24. 7,807
    − 4,383

25. 4,510
    − 4,238

26. 7,030
    − 5,426

27. 3,498
    − 2,499

28. 87,093
    − 54,321

29. 67,293
    − 34,089

30. 12,786
    − 7,398

Section 1-4 Multiplication of Whole Numbers

The operation of multiplication of whole numbers is often referred to as "shortcut addition." The multiplication

$$2 \times 3 = 6$$

can be thought of as the process used to obtain two 3s. The same result could be obtained by adding two 3s. This explains the idea of shortcut addition.

In the problem $2 \times 3 = 6$, the two numbers that are being multiplied are called **factors**. The answer obtained is called the **product**.

$$2 \times 3 = 6$$
factors   product

The Distributive Law   The distributive law provides an important link between multiplication and addition.

Distributive Law

For any number a, b, and c,
$a \times (b + c) = a \times b + a \times c$

Have you used this law? You have used this law when multiplying a two-digit number by a single-digit number. The arrows below indicate that each digit of 23 is multiplied by 3:

Problem:

$$\begin{array}{r} 23 \\ \times\ 3 \\ \hline 69 \end{array}$$

Reasoning:

$3 \times 23 = 3 \times (20 + 3)$
$\quad\quad\quad = 3 \times 20 + 3 \times 3$ **distributive law**
$\quad\quad\quad = 60 + 9$
$\quad\quad\quad = 69$

(*Note:* It is really 20 that is being multiplied by 3.)

The next example shows a case in which a number is carried.

$$\begin{array}{r} \overset{2}{} \\ 28 \\ \times\ 3 \\ \hline 84 \end{array}$$

$3 \times 8 = 24 = 20 + 4$
write 4
carry 2 (really 2 tens)

As in addition problems, one uses the associative law of addition when carrying a number.

In the problem below, do you know why the 64 is placed as it is below the 96?

Problem: *Reasoning:*

```
    32                  23 × 32 = (20 + 3) × 32
   ×23                         = 20 × 32 + 3 × 32    distributive law
    96    ← 3 × 32              = 640 + 96
    64    ← 2 × 32
   ———
   736      ↑
          digits of 23
```

Note that 32 is really being multiplied by 20. The problem could have been written as shown below.

```
    32
   ×23
    ——
    96    ← 3 × 32
   640    ← 20 × 32
   ———
   736
```

The example above suggests a simple procedure for multiplying 327 × 496. Multiply 496 by each "part" of 327; that is, by 300, by 20, and by 7. This is demonstrated in the following sample problem.

Sample Problem 1-7

Solution:

Multiply 327 × 496.

Since 327 = 300 + 20 + 7, multiply 496 by 300, by 20, and by 7:

```
       496
      ×327
     —————
     3,472    ←   7 × 496
     9,920    ←  20 × 496
   148,800    ← 300 × 496
   ———————
   161,192
```

When you become familiar with the process just used, you may prefer not to write the extra zeros.

```
       496
      ×327
     —————
     3 472    ← 7 × 496
     9 92     ← 2 × 496
   148 8      ← 3 × 496
   ———————         ↑
   161,192    the digits of 327
```

Complete the problems in the sample set. Completed problems appear on page 24.

Sec. 1-4 Multiplication of Whole Numbers

Sample Set

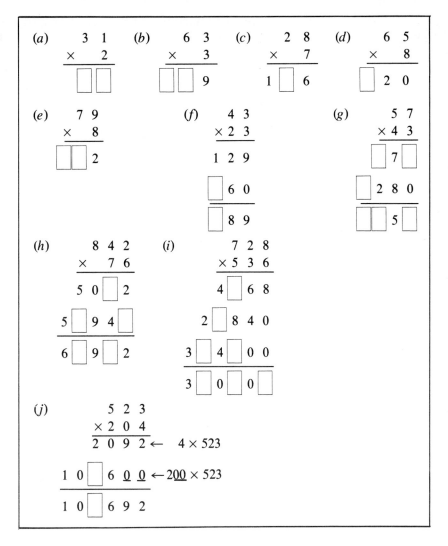

The Commutative and Associative Laws of Multiplication

The commutative and associative laws of multiplication are stated below:

Commutative Law of Multiplication

For any numbers a and b,
$$a \times b = b \times a$$

Example $5 \times 7 = 35$ and $7 \times 5 = 35$

Associative Law of Multiplication

For any numbers a, b, and c,
$$a \times (b \times c) = (a \times b) \times c$$

Example $3 \times (4 \times 2) = 3 \times 8 = 24$
$(3 \times 4) \times 2 = 12 \times 2 = 24$

24 Whole-Number Arithmetic

The commutative and associative laws of multiplication tell us that numbers in a product may be rearranged and multiplied in any order. This fact will be used later.

Completed Problems

(a) 31
  × 2
  ─────
  6 [2]

(b) 63
  × 3
  ─────
  1 [8] 9

(c) 28
  × 7
  ─────
  1 [9] 6

(d) 65
  × 8
  ─────
  [5] 2 0

(e) 79
  × 8
  ─────
  6 [3] 2

(f) 43
  ×23
  ─────
  1 2 9
  [8] 6 0
  ─────
  [9] 8 9

(g) 57
  ×43
  ─────
  1 7 [1]
  [2] 2 8 0
  ─────
  [2] [4] 5 [1]

(h) 842
  × 76
  ─────
  5 0 [5] 2
  5 [8] 9 4 [0]
  ─────
  6 [3] 9 [9] 2

(i) 728
  ×536
  ─────
  4 [3] 6 8
  2 [1] 8 4 0
  3 [6] 4 [0] 0 0
  ─────
  3 [9] 0 [2] 0 [8]

(j) 523
  ×204
  ─────
  2 0 9 2 ← 4 × 523
  1 0 [4] 6 0 0 ← 2̲0̲0̲ × 523
  ─────
  1 0 [6] 6 9 2

Name: _____

Class: _____

Exercises for Sec. 1-4 Find the products.

1. 12   2. 21   3. 32   4. 83
   × 4   × 6   × 3   × 7

5. 38   6. 47   7. 76   8. 32
   × 9   × 6   × 9   ×15

9. 54   10. 73   11. 67   12. 94
   ×32   ×20   ×36   ×84

13. 78   14. 97   15. 139   16. 233
   ×57   ×46   × 21   × 46

17. 532   18. 671   19. 502   20. 804
   × 65   × 86   × 24   × 87

21. 1843   22. 3454   23. 5267   24. 8209
   × 32   × 56   × 95   × 29

25. 7286
   × 97

26. 1524
   × 321

27. 5674
   × 508

28. 7369
   × 632

29. 2637
   × 385

30. 6958
   × 769

Section 1-5 Division of Whole Numbers

Division and multiplication are inverse operations. In order to see this, observe the two problems below:

$$3 \times 4 = 12$$
and
$$12 \div 3 = 4$$

The division operation in the second problem can be thought of as "undoing" the multiplication operation in the first. Starting with 4 in the first problem, a multiplication operation was performed to arrive at 12. In the second, a division operation was used in order to get back to 4. In this sense, the operations are inverse.

Each of the numbers in a division problem has a special name:

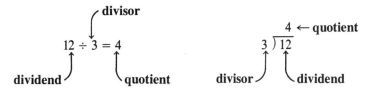

The fact that multiplication and division are inverse operations provides a method for checking division. If the division is correct, the product of the divisor and quotient must equal the dividend.

Problem $12 \div 3 = 4$ dividend $\div$ divisor = quotient
Check: $3 \times 4 = 12$ divisor $\times$ quotient = dividend

Note the definitions given below:

If $c \div a = b$, and b is a whole number, then c is **divisible** by a.
If $c = b \times a$, and b is a whole number, then c is a **multiple** of a.

Example   Since $12 \div 3 = 4$ ← a whole number
                                  12 is divisible by 3
        Since $12 = 4 \times 3$ a whole number
                                  12 is a multiple of 3

We can conclude that a first number is divisible by a second if the first is a multiple of the second.

Very likely you have viewed a division problem as asking how many times one number will "go into" another. For example, the problem

$$12 \div 3$$

can be thought of as asking the question: "How many 3s are there in 12?" Since

$$12 \div 3 = 4 \quad \text{whole number}$$

3 goes into 12 exactly 4 times with no remainder.

28 Whole-Number Arithmetic

In the problem

$$14 \div 3$$

the dividend 14 is not divisible by 3. Think of the problem as asking: "How many 3s are there in 14?" The equality

$$14 = 4 \times 3 + 2 \qquad \text{multiplication before addition}$$

tells us that 14 is composed of four 3s with 2 left over, or 3 goes into 14 four times with remainder 2.

In familiar notation, this problem can be written

$$\begin{array}{r} 4 \\ 3\overline{)14} \\ \underline{12} = 4 \times 3 \\ 2 \ \text{remainder} \end{array} \qquad \text{or better} \qquad \begin{array}{r} 4\tfrac{2}{3} \\ 3\overline{)14} \end{array} \qquad (\textit{Note: } 4\tfrac{2}{3} = 4 + \tfrac{2}{3})$$

The number 12 is the largest multiple of 3 that is less than 14. The multiplier 4 used to obtain this multiple becomes the **whole-number part** of the quotient. The **fractional part** of the quotient is formed by placing the remainder over the divisor. This will be justified at the end of the section.

Division can be checked by adding the remainder to the product of the whole-number part of the quotient and divisor. If the division is correct, the result will be the dividend.

Check: $\qquad\qquad\qquad 4 \times 3 + 2 = 12 + 2 = 14$
Whole part of quotient $\times$ divisor $+$ remainder $=$ dividend

Long Division The long-division process gives us the whole-number part of the quotient. When this number is multiplied by the divisor, the product is the largest multiple of the divisor equal to or less than the dividend. In order to find the whole part of the quotient, one could compute many multiples until the desired one is found. This process could be very time-consuming. In the sample problems that follow, this task is done by using repeated short divisions and educated guesses.

Sample Problem 1-8 Divide $853 \div 6$.

Solution:

Step 1. Since 6 is less than 8, 6 goes into 8 once.
(a) Write 1 above the 8:
(b) Subtract $8 - 6 = 2$:
(c) Bring down 5:

$$\begin{array}{r} 1 \\ 6\overline{)853} \\ \underline{-6}\downarrow \\ 25 \end{array} \qquad \text{bring down 5}$$

Step 2. How many 6s in 25?
Correct guess $= 4$
(a) Write 4 above the 5:
(b) Subtract $24 \ (= 4 \times 6)$ from 25:
(c) Bring down 3:

$$\begin{array}{r} 14 \\ 6\overline{)853} \\ \underline{6} \\ 25 \\ \underline{24}\downarrow \\ 13 \end{array} \qquad \text{bring down 3}$$

Sec. 1-5 Division of Whole Numbers

> Step 3. How many 6s in 13?
> Correct guess = 2
> (a) Write 2 above the 3:
> (b) Subtract 12 (= 2 × 6) from 13:
> (c) Write the remainder 1 over the divisor 6 to form the fractional part of the quotient:
>
> $$\begin{array}{r} 142\frac{1}{6} \\ 6\overline{)853} \\ 6 \\ \hline 25 \\ 24 \\ \hline 13 \\ 12 \\ \hline 1 \end{array}$$
>
> Check: Whole part × divisor + remainder = dividend
> 142 × 6 + 1 = 852 + 1 = 853

The problem just worked is written below with zeros filled in and the 3 brought down in the second step. Notice that in each step, a multiple of 6 is subtracted until a remainder less than 6 is obtained. Will the sum of these multiples yield the largest multiple of 6 that is less than 853? The multiplication on the right indicates this is the case.

Problem

$$\begin{array}{r} 142 \\ 6\overline{)853} \\ -600 \\ \hline 253 \\ -240 \\ \hline 13 \\ -12 \\ \hline 1 \end{array}$$
  ← 100 × 6
  ← 40 × 6
  ← 2 × 6

multiples of 6

$$\begin{array}{r} 6 \\ \times 142 \\ \hline 12 \\ 240 \\ 600 \\ \hline 852 \end{array}$$
  ← divisor
  ← whole part of quotient
  ← 2 × 6
  ← 40 × 6
  ← 100 × 6
  ← largest multiple of the divisor less than 853

Sample Problem 1-9   Divide 583 ÷ 8.

Solution:

> Step 1. Since 8 is greater than 5, how many 8s in 58?
> Correct guess = 7
> (a) Write 7 above the 8:
> (b) Subtract 56 (= 7 × 8) from 58:
> (c) Bring down 3:
>
> $$\begin{array}{r} 7 \\ 8\overline{)583} \\ 56 \\ \hline 23 \end{array}$$
>
> Step 2. How many 8s in 23?
> Correct guess = 2
> (a) Write 2 above the 3:
> (b) Subtract 16 (= 2 × 8) from 23:
> (c) Form the fractional part of the quotient:
>
> $$\begin{array}{r} 72\frac{7}{8} \\ 8\overline{)583} \\ 56 \\ \hline 23 \\ 16 \\ \hline 7 \end{array}$$
>
> Check: 72 × 8 + 7 = 576 + 7 = 583

Sample Problem 1-10

Solution:

Divide 14525 ÷ 72.

Step 1.   Since 72 is greater than 1 or 14, how many 72s in 145?
Since 140 ÷ 70 = 2, we might guess that there are **two** 72s in 145. This guess is correct.
(a) Write 2 above the 5:
(b) Subtract 144 (= 2 × 72) from 145:
(c) Bring down 2:

$$\begin{array}{r} 2 \\ 72\overline{)14525} \\ 144 \\ \hline 12 \end{array}$$

Step 2.   Since 72 will not go into 12:
(a) Place 0 above the 2:
(b) Bring down 5:

$$\begin{array}{r} 20 \\ 72\overline{)14525} \\ 144 \\ \hline 125 \end{array}$$

Step 3.   How many 72s in 125?
Correct guess = 1
(a) Write 1 above the 5:
(b) Subtract 72 (= 1 × 72) from 125:
(c) Form the fractional part of the quotient:

$$\begin{array}{r} 201\tfrac{53}{72} \\ 72\overline{)14525} \\ 144 \\ \hline 125 \\ 72 \\ \hline 53 \end{array}$$

Check: 201 × 72 + 53 = 14,472 + 53 = 14,525

Complete the problems below. Completed problems appear on page 32.

Sample Set

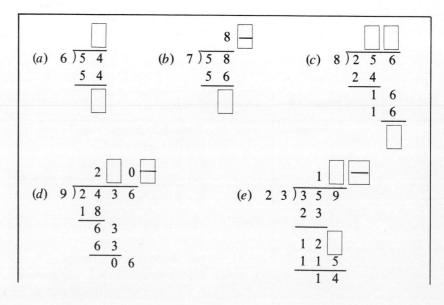

```
            6 ☐ ─                                3 ☐☐
 (f) 5 7 ) 3 8 4 6                   (g) 2 4 ) 7 3 2 0
           3 4 2                                 7 2
           ─────                                 ─────
             4 2 6                                 1 2 0
             3 9 9                                 1 2 0
             ─────                                 ─────
                ☐☐                                     0
```

Division Check Earlier in this section, it was stated that the fractional part of a quotient is obtained by placing the remainder over the divisor.

$$\begin{array}{r} 4\frac{2}{3} \\ 3\overline{)14} \\ \underline{12} \\ 2 \end{array}$$

(*Note:* $4\frac{2}{3} = 4 + \frac{2}{3}$)

This will be justified by showing that $4\frac{2}{3}$ is the entire quotient. Since multiplication and division are inverse operations, this is equivalent to showing

$$3 \times 4\frac{2}{3} = 14$$

This is easily seen using the distributive law.

$$\begin{aligned} 3 \times 4\tfrac{2}{3} &= 3 \times (4 + \tfrac{2}{3}) \\ &= 3 \times 4 + 3 \times \tfrac{2}{3} \quad &&\text{distributive law} \\ &= 12 + 2 \quad &&\tfrac{2}{3} \text{ of 3 is 2} \\ &= 14 \end{aligned}$$

Laws of Mathematics

Commutative law of addition	$a + b = b + a$
Associative law of addition	$a + (b + c) = (a + b) + c$
Commutative law of multiplication	$a \times b = b \times a$
Associative law of multiplication	$a \times (b \times c) = (a \times b) \times c$
Distributive law	$a \times (b + c) = a \times b + a \times c$

Completed Problems

(a)
```
       9
   6)5 4
     5 4
       0
```

(b)
```
       8 2/7
   7)5 8
     5 6
       2
```

(c)
```
        3 2
   8)2 5 6
     2 4
       1 6
       1 6
         0
```

(d)
```
       2 7 0 6/9
   9)2 4 3 6
     1 8
       6 3
       6 3
         0 6
```

(e)
```
         1 5 14/23
   2 3)3 5 9
       2 3
         1 2 9
         1 1 5
             1 4
```

(f)
```
           6 7 27/57
   5 7)3 8 4 6
       3 4 2
           4 2 6
           3 9 9
               2 7
```

(g)
```
           3 0 5
   2 4)7 3 2 0
       7 2
           1 2 0
           1 2 0
               0
```

Name: _____

Class: _____

Exercises for Sec. 1-5 Find each quotient.

1. 5)75
2. 4)852
3. 6)552

4. 7)658
5. 4)824
6. 9)734

7. 6)875
8. 12)756
9. 15)345

10. 13)585
11. 27)594
12. 35)945

13. 56)1,792
14. 37)1,591
15. 83)4,233

16. 47)2,036
17. 67)3,358
18. 52)1,903

19. 96)6,773
20. 73)15,549
21. 46)15,778

22. $28 \overline{)14{,}196}$ 23. $56 \overline{)40{,}880}$ 24. $86 \overline{)27{,}319}$

25. $47 \overline{)40{,}067}$ 26. $327 \overline{)8{,}156}$ 27. $536 \overline{)13{,}400}$

28. $214 \overline{)12{,}891}$ 29. $873 \overline{)46{,}310}$ 30. $753 \overline{)457{,}384}$

Name: _____

Class: _____

Unit 1 Test Write the numbers in expanded form.

1. 96 =

2. 729 =

3. 8,032 =

4. 43,965 =

Find the sums.

5. 38
   29
 +54

6. 803
   892
 +379

7. 76
  35
  58
 +47

8. 572
   461
   783
 +715

Find the differences.

9. 72
 −56

10. 305
  − 79

11. 574
 −258

12. 8,752
 −8,692

Find the products.

13. 38
 × 9

14. 72
 ×53

15. 369
 × 78

16. 498
 ×267

Find the quotients.

17. 6)87

18. 7)534

19. 37)2,965

20. 84)27,185

Section 2-1 Prime Factorization

It was stated earlier that parentheses are used to indicate the order in which operations are to be performed. Parentheses can also be used to indicate multiplication. For example,
$$(3)(4) = 3 \times 4 = 12$$

Whole numbers that form a product are called **factors** of that product.

$12 = (3)\,(4)$ — factors of 12

$12 = (2)\,(6)$ — factors of 12

$12 = (1)\,(12)$ — factors of 12

Since each factor in a product must be a whole number, the product must be divisible by both.

$12 \div 3 = 4$ or $12 \div 4 = 3$

$12 \div 2 = 6$ or $12 \div 6 = 2$

$12 \div 1 = 12$ or $12 \div 12 = 1$

Factors of a given number are those whole numbers by which the given number is divisible.

Factors of 12:   1,   2,   3,   4,   6,   12
Factors of 16:   1,   2,   4,   8,   16
Factors of 30:   1,   2,   3,   5,   6,   10,   15,   30

A number whose only factors are 1 and the number itself is a **prime number**. The prime numbers less than 100 are listed below:

2,   3,   5,   7,   11
13,   17,   19,   23,   29
31,   37,   41,   43,   47
53,   59,   61,   67,   71
73,   79,   83,   89,   97

The number 1 is usually not included in the list.

Mathematicians have shown that there is no largest prime number.

A whole number that is not a prime number is called a **composite number**.

A whole number is **factored** when written as a product of factors. This product is called a **factorization**. If the factors are all **prime factors** (prime numbers), the factorization is called a **prime factorization**.

$54 = (6)\,(9)$ **factorization of 54**

$54 = (2)\,(3)\,(3)\,(3)$ **prime factorization of 54**

40 Prime Factorization, Equivalent Fractions, Least Common Multiple

> **There is only one prime factorization of a given number.**

In the prime factorization, it is customary to write the primes in ascending order—that is, smallest primes first, and largest last. A method for finding the prime factorization of a number is demonstrated below.

Sample Problem 2-1 Write the prime factorization of 54.

Solution:

Step 1. Divide 54 by 2 (the smallest prime):

$$2 \overline{)54} \\ \overline{27}$$

Step 2. Since 2 is not a prime factor of 27, divide by 3 (the next prime after 2):

$$2 \overline{)54} \\ 3 \overline{)27} \\ \overline{9}$$

Step 3. Divide 9 by 3. The divisors used in each step, together with the final quotient, are the prime factors of 54:
$54 = (2)(3)(3)(3)$

$$2 \overline{)54} \\ 3 \overline{)27} \\ 3 \overline{)9} \\ \overline{3}$$

Sample Problem 2-2 Write the prime factorization of 144.

Solution:

Step 1. Divide 144 by 2:

$$2 \overline{)144} \\ \overline{72}$$

Step 2. Since 2 is a prime factor of 72, divide by 2:

$$2 \overline{)144} \\ 2 \overline{)72} \\ \overline{36}$$

Step 3. Since 2 is a prime factor of 36, divide by 2:

$$2 \overline{)144} \\ 2 \overline{)72} \\ 2 \overline{)36} \\ \overline{18}$$

Step 4. Since 2 is a prime factor of 18, divide by 2:

$$2 \overline{)144} \\ 2 \overline{)72} \\ 2 \overline{)36} \\ 2 \overline{)18} \\ \overline{9}$$

Sec. 2-1 Prime Factorization

Step 5. 2 is not a prime factor of 9; divide by 3 (a prime factor of 9):

```
2 ) 144
2 ) 72
2 ) 36
2 ) 18
3 ) 9
    3
```

Since the final quotient is a prime, we are finished:

$$144 = (2)(2)(2)(2)(3)(3)$$

Complete the factorizations below. Completed factorizations appear on page 42.

Sample Set

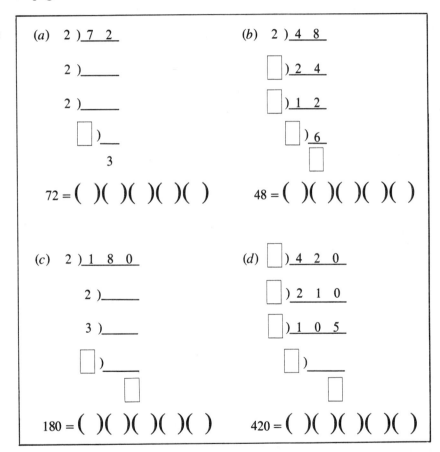

(a)
```
2 ) 7 2
2 )___
2 )___
 ) ___
   3
```
72 = ( )( )( )( )( )

(b)
```
2 ) 4 8
 ) 2 4
 ) 1 2
 ) 6
```
48 = ( )( )( )( )( )

(c)
```
2 ) 1 8 0
2 )___
3 )___
 )___
```
180 = ( )( )( )( )( )

(d)
```
 ) 4 2 0
 ) 2 1 0
 ) 1 0 5
 )___
```
420 = ( )( )( )( )( )

When factoring numbers, it is helpful to know divisibility rules for the prime factors 2, 3, and 5. These rules are listed below.

Divisibility Rules

1 **A number ending in 0, 2, 4, 6, or 8 is divisible by 2.** Such numbers are called *even* numbers.

2 **If the sum of the digits of a number is divisible by 3, the number itself is divisible by 3.**
Example:
The sum of the digits of 87 is 8 + 7 = 15. Since 15 is divisible by 3, 87 is divisible by 3.

3 **A number ending in 0 or 5 is divisible by 5.**

Sample Problem 2-3

Write the prime factorization of 630.

Solution:

$$\begin{array}{r} 2\,)\,\overline{630} \\ 3\,)\,\overline{315} \\ 3\,)\,\overline{105} \\ 5\,)\,\overline{35} \\ 7 \end{array}$$

← even   divide by 2
← 3 + 1 + 5 = 9   divide by 3
← 1 + 0 + 5 = 6   divide by 3
← ends in 5   divide by 5

630 = (2) (3) (3) (5) (7)

Completed Factorizations

(a)
$$\begin{array}{r} 2\,)\,\overline{7\ 2} \\ 2\,)\,\overline{36} \\ 2\,)\,\overline{18} \\ 3\,)\,\overline{9} \\ 3 \end{array}$$

72 = (2)(2)(2)(3)(3)

(b)
$$\begin{array}{r} 2\,)\,\overline{4\ 8} \\ 2\,)\,\overline{2\ 4} \\ 2\,)\,\overline{1\ 2} \\ 2\,)\,\overline{6} \\ 3 \end{array}$$

48 = (2)(2)(2)(2)(3)

(c)
$$\begin{array}{r} 2\,)\,\overline{1\ 8\ 0} \\ 2\,)\,\overline{90} \\ 3\,)\,\overline{45} \\ 3\,)\,\overline{15} \\ 5 \end{array}$$

180 = (2)(2)(3)(3)(5)

(d)
$$\begin{array}{r} 2\,)\,\overline{4\ 2\ 0} \\ 2\,)\,\overline{2\ 1\ 0} \\ 3\,)\,\overline{1\ 0\ 5} \\ 5\,)\,\overline{35} \\ 7 \end{array}$$

420 = (2)(2)(3)(5)(7)

Write the prime factorization of each product in the multiplication table below. Then try doing the work entirely in your head. You will be amazed how quickly and thoroughly you learn the products.

×	1	2	3	4	5	6	7	8	9
1	1	2	3	4	5	6	7	8	9
2	2	4	6	8	10	12	14	16	18
3	3	6	9	12	15	18	21	24	27
4	4	8	12	16	20	24	28	32	36
5	5	10	15	20	25	30	35	40	45
6	6	12	18	24	30	36	42	48	54
7	7	14	21	28	35	42	49	56	63
8	8	16	24	32	40	48	56	64	72
9	9	18	27	36	45	54	63	72	81

Name: _____

Class: _____

Exercises for Sec. 2-1 Write the prime factorization of each number.

1. 6 =

2. 9 =

3. 8 =

4. 15 =

5. 16 =

6. 18 =

7. 20 =

8. 21 =

9. 24 =

10. 25 =

11. 28 =

12. 38 =

Prime Factorization, Equivalent Fractions, Least Common Multiple

13. 40 =

14. 56 =

15. 60 =

16. 64 =

17. 70 =

18. 72 =

19. 80 =

20. 81 =

21. 84 =

22. 96 =

23. 100 =   24. 108 =

25. 121 =   26. 128 =

27. 152 =   28. 160 =

29. 180 =   30. 200 =

31. 216 =   32. 256 =

33. 288 =   34. 243 =

35. 360 =   36. 432 =

37. 640 =   38. 864 =

39. 1600 =   40. 8100 =

Section 2-2 Reducing Fractions

A common fraction is a fraction of the form $\frac{a}{b}$ (also written a/b), where a and b are whole numbers. The number a is the **numerator,** and b (not equal to 0) is the **denominator.**

The fraction $1/b$ can be thought of as the part of 1 obtained when 1 whole object is divided into b equal parts.

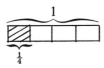
1 rectangle divided into 4 equal parts; each part is $\frac{1}{4}$ of 1 rectangle

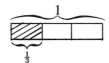
1 rectangle divided into 3 equal parts; each part is $\frac{1}{3}$ of 1 rectangle

The fraction a/b can be interpreted as the a of the parts $1/b$.

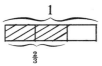
$\frac{2}{3} = 2 \times \frac{1}{3} = 2$ of the thirds; $\frac{2}{3}$ of 1 whole rectangle

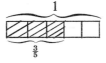
$\frac{3}{5} = 3 \times \frac{1}{5} = 3$ of the fifths; $\frac{3}{5}$ of 1 whole rectangle

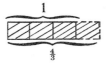
$\frac{4}{3} = 4 \times \frac{1}{3} = 4$ of the thirds; more than 1 whole rectangle

> The fraction $\frac{a}{b}$ is defined to be $a \times \frac{1}{b}$

The fraction $b/b = 1$.

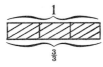
$\frac{3}{3} = 3 \times \frac{1}{3} = 3$ of the thirds; same as 1 whole rectangle

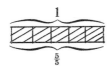
$\frac{5}{5} = 5 \times \frac{1}{5} = 5$ of the fifths; same as 1 whole rectangle

Fractions in which the numerator is less than the denominator are called **proper fractions**. Thus,

$$\tfrac{1}{2} \quad \tfrac{2}{5} \quad \tfrac{8}{9} \quad \tfrac{39}{47} \quad \text{are proper fractions}$$

A fraction in which the numerator is greater than or equal to the denominator is an **improper fraction**. Thus,

$$\tfrac{3}{2} \quad \tfrac{9}{5} \quad \tfrac{5}{5} \quad \tfrac{59}{47} \quad \text{are improper fractions}$$

Equivalent Fractions

Two fractions may express the same part of an object. Such fractions are **equivalent**. Consider the following:

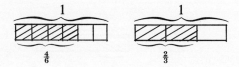

Here the fractions $\tfrac{4}{6}$ and $\tfrac{2}{3}$ represent the same part of a rectangle, or the same number. They are therefore equivalent.

A better understanding of equivalent fractions depends on the following properties.

1 **For any number N,**

$$N \times 1 = 1 \times N = N$$

2 **Multiplication rule for fractions:**

$$\frac{a}{b} \times \frac{c}{d} = \frac{a \times c}{b \times d}$$

The numerator and denominator of $\tfrac{4}{6}$ have a common factor 2.

$$\frac{4}{6} = \frac{2 \times 2}{3 \times 2}$$

$$= \frac{2}{3} \times \frac{2}{2} \qquad \textbf{multiplication rule}$$

$$= \frac{2}{3} \times 1 \qquad \frac{2}{2} = 1$$

$$= \frac{2}{3} \qquad N \times 1 = N$$

A fraction is **reduced** when it is written as an equivalent fraction with smaller denominator. Above, $\tfrac{4}{6}$ was reduced to $\tfrac{2}{3}$. Another fraction that reduces to $\tfrac{2}{3}$ is the fraction $\tfrac{6}{9}$.

$$\frac{6}{9} = \frac{2 \times 3}{3 \times 3} = \frac{2}{3} \times \frac{3}{3} = \frac{2}{3} \times 1 = \frac{2}{3}$$

Sec. 2-2 Reducing Fractions

> Two fractions are equivalent if one fraction reduces to the other or if both reduce to the same fraction.

The fraction $\frac{6}{9}$, $\frac{4}{6}$, and $\frac{2}{3}$ are all equivalent.

$$\frac{6}{9} = \frac{4}{6} = \frac{2}{3}$$

It is not necessary to reduce fractions in order to see that they are equivalent.

$$\frac{a}{b} = \frac{c}{d} \quad \text{if} \quad a \times d = b \times c$$

The products $a \times d$ and $b \times c$ are called **cross products**. Note that $\frac{6}{9}$ and $\frac{4}{6}$ satisfy the above requirement.

$$\frac{6}{9} = \frac{4}{6} \quad \text{since} \quad 6 \times 6 = 9 \times 4$$

When a fraction is reduced to the equivalent fraction with smallest possible denominator, it is **reduced to lowest terms**.

Sample Problem 2-4

Solution:

Reduce $\frac{18}{12}$ to lowest terms.

Step 1. Write the numerator and denominator as products of prime factors:

$$\frac{18}{12} = \frac{(2)(3)(3)}{(2)(2)(3)}$$

Step 2. Forming fractions equal to 1, write the fraction as a product:

$$= \frac{2}{2} \times \frac{3}{2} \times \frac{3}{3}$$

$$= 1 \times \frac{3}{2} \times 1$$

The commutative and associative laws of multiplication were used to rearrange the numbers.

$$= \frac{3}{2}$$

Read on; better things are yet to come.

The process used to reduce fractions can be simplified using **cancellation** of common factors in numerator and denominator.

Sample Problem 2-5 Reduce $\frac{18}{12}$ to lowest terms.

Solution:

Step 1. Write the numerator and denominator as products of prime factors:
$$\frac{18}{12} = \frac{(\cancel{2})(3)(\cancel{3})}{(\cancel{2})(2)(\cancel{3})}$$

Step 2. Cancel a 2 in the numerator and denominator; cancel a 3 in the numerator and denominator:
$$= \frac{3}{2}$$

(*Note:* Since there was only one 3 in the denominator, only one 3 can be canceled in the numerator.)

Read on; better things are yet to come.

Cancellation is not a true mathematical operation. It is simply a handy shortcut dependent on recognizing that $\frac{2}{2} = 1$, and $\frac{3}{3} = 1$, and

$$\frac{(2)(3)(3)}{(2)(2)(3)} = \frac{2}{2} \times \frac{3}{2} \times \frac{3}{3} = 1 \times \frac{3}{2} \times 1 = \frac{3}{2}$$

A common factor of two numbers is a number that is a factor of both.

A fraction can be reduced if the numerator and denominator have a common factor.

Dividing both numerator and denominator by a common factor is equivalent to canceling the factor.

$$\frac{\cancel{4}^2}{\cancel{6}_3} = \frac{(\cancel{2})(2)}{(\cancel{2})(3)} = \frac{2}{3} \qquad \text{divide numerator and denominator by 2}$$

$$\frac{\cancel{18}^3}{\cancel{12}_2} = \frac{(\cancel{6})(3)}{(\cancel{6})(2)} = \frac{3}{2} \qquad \text{divide numerator and denominator by 6}$$

Note that the canceled factor need not be a prime. Dividing the numerator and denominator by their largest common factor shortens the reduction process. If it is not apparent what the largest common factor is, the cancellation must be done in steps.

Sample Problem 2-6 Reduce $\frac{144}{162}$ to lowest terms.

Solution:

Step 1. Since both the numerator and denominator are even, divide by 2:
$$\begin{array}{r} 72 \\ \cancel{144} \\ \cancel{162} \\ 81 \end{array}$$

53 Sec. 2-2 Reducing Fractions

> Step 2. Since 72 and 81 are divisible by 9, divide by 9 (using 3 would be slow):
>
> Since 8 and 9 have no common factor, the fraction is reduced to lowest terms.

$$\frac{\cancel{144}^{\,\cancel{72}^{\,8}}}{\cancel{162}_{\,\cancel{81}_{\,9}}} = \frac{8}{9}$$

Shortcuts are fine. However, try not to lose sight of the principles involved.

Complete each problem below. Completed problems appear on page 54.

Sample Set

(a) $\dfrac{4}{6} = \dfrac{\Box}{3}$

(b) $\dfrac{\cancel{8}}{\cancel{20}_{5}} = \dfrac{2}{\Box}^{\Box}$

(c) $\dfrac{\cancel{14}^{7}}{\cancel{10}_{\Box}} = \dfrac{\Box}{\Box}$

(d) $\dfrac{\cancel{25}}{\cancel{10}}^{\Box}_{\Box} = \dfrac{5}{2}$

(e) $\dfrac{\cancel{24}^{\cancel{\Box}}}{\cancel{36}_{\cancel{\Box}_{3}}} = \dfrac{\Box}{3}$

(f) $\dfrac{\cancel{72}^{\cancel{36}^{\cancel{\cancel{8}}^{3}}}}{\cancel{96}_{\cancel{\Box}_{\cancel{\Box}_{\Box}}}} = \dfrac{\Box}{\Box}$

A Very Important Point

When is a fraction equivalent to a whole number?

$$\tfrac{24}{6} = 4$$

A fraction reduces to a whole number if

1. **Every prime factor of the denominator appears as a prime factor of the numerator.**

$$\frac{24}{6} = \frac{(2)(2)(\cancel{2})^{1}(\cancel{3})^{1}}{(\cancel{2})_{1}(\cancel{3})_{1}} = 4$$

2. **The numerator is divisible by the denominator.**

$$24 \div 6 = 4$$

3. **The numerator is a multiple of the denominator.**

$$24 = (4)(6)$$

54 Prime Factorization, Equivalent Fractions, Least Common Multiple

Quotient = Fraction   The following definition will be found in advanced mathematics books.

Definition

$$a \div b = \frac{a}{b}$$

This implies that the quotient is simply the fraction formed by placing the dividend over the divisor. As examples,

$$6 \div 2 = \tfrac{6}{2} \quad \textbf{six-halves}$$
$$1 \div 3 = \tfrac{1}{3} \quad \textbf{one-third}$$

In the first example, the definition yields the familiar answer 3 since the fraction $\tfrac{6}{2}$ reduces to 3.

The second example may be viewed as dividing 1 whole unit into 3 equal parts. Each resulting part would then be equal to $\tfrac{1}{3}$ of a unit.

At the beginning of this section, it was stated that the denominator of a fraction cannot be 0. Placing 0 in the denominator would be equivalent to dividing by 0. For example,

$$\tfrac{2}{0} = 2 \div 0$$

Can this division problem have an answer? Let b represent a possible answer. Since division and multiplication are inverse operations, we must have **both**

$$2 \div 0 = b \quad \text{and} \quad 2 = 0 \times b \quad \textbf{impossible}$$

The latter is an impossible product since $0 \times b = 0$, not 2. Since the multiplication is impossible, the division is also impossible. For this reason we say that division by zero is not defined.

Completed Problems

(a) $\dfrac{4}{6} = \dfrac{2}{3}$

(b) $\dfrac{\overset{2}{\cancel{8}}}{\underset{5}{\cancel{20}}} = \dfrac{2}{5}$

(c) $\dfrac{\overset{7}{\cancel{14}}}{\underset{5}{\cancel{10}}} = \dfrac{7}{5}$

(d) $\dfrac{\overset{5}{\cancel{25}}}{\underset{2}{\cancel{10}}} = \dfrac{5}{2}$

(e) $\dfrac{\overset{\overset{2}{\cancel{4}}}{\cancel{24}}}{\underset{\underset{3}{\cancel{6}}}{\cancel{36}}} = \dfrac{2}{3}$

(f) $\dfrac{\overset{\overset{\overset{3}{\cancel{6}}}{\cancel{36}}}{\cancel{72}}}{\underset{\underset{\underset{4}{\cancel{8}}}{\cancel{48}}}{\cancel{96}}} = \dfrac{3}{4}$

Name: _____

Class: _____

Exercises for Sec. 2-2 Reduce the fractions to lowest terms.

1. $\dfrac{6}{9} =$ 　　　2. $\dfrac{8}{12} =$ 　　　3. $\dfrac{9}{15} =$

4. $\dfrac{16}{8} =$ 　　　5. $\dfrac{10}{12} =$ 　　　6. $\dfrac{12}{9} =$

7. $\dfrac{18}{12} =$ 　　　8. $\dfrac{10}{15} =$ 　　　9. $\dfrac{24}{36} =$

10. $\dfrac{28}{24} =$ 　　　11. $\dfrac{18}{30} =$ 　　　12. $\dfrac{34}{30} =$

13. $\dfrac{16}{32} =$ 　　　14. $\dfrac{25}{30} =$ 　　　15. $\dfrac{54}{36} =$

16. $\dfrac{24}{56} =$ 　　　17. $\dfrac{72}{56} =$ 　　　18. $\dfrac{81}{72} =$

19. $\dfrac{84}{96} =$ 　　　20. $\dfrac{96}{84} =$ 　　　21. $\dfrac{144}{72} =$

22. $\dfrac{81}{144} =$ 23. $\dfrac{100}{144} =$ 24. $\dfrac{96}{144} =$

25. $\dfrac{128}{152} =$ 26. $\dfrac{128}{288} =$ 27. $\dfrac{189}{294} =$

28. $\dfrac{300}{360} =$ 29. $\dfrac{210}{252} =$ 30. $\dfrac{256}{560} =$

Section 2-3 Equivalent Fractions with Larger Denominator

It is often necessary to write a fraction as an equivalent fraction with a larger denominator. To do this, we multiply the fraction by a fraction of the form c/c (= 1). Thus,

$$\frac{a}{b} = \frac{a}{b} \times 1$$

$$= \frac{a}{b} \times \frac{c}{c}$$

$$= \frac{a \times c}{b \times c} \quad \text{a fraction equivalent to } \frac{a}{b}$$

Notice that this procedure is exactly the opposite of the one used to reduce fractions.

$$\frac{2}{3} \times \frac{2}{2} = \frac{4}{6}$$
$$\frac{2}{3} \times \frac{3}{3} = \frac{6}{9} \longleftarrow \text{fractions equivalent to } \frac{2}{3}$$
$$\frac{2}{3} \times \frac{4}{4} = \frac{8}{12}$$

When a new denominator is given, the task is to find the numerator that will make the fractions equivalent. For example, we wish to express $\frac{2}{3}$ as a number of 24th.

$$\frac{2}{3} \times \frac{b}{b} = \frac{?}{24}$$

The denominator 24 is the product of the denominators 3 and b.

$$3 \times b = 24$$

Since division is the inverse operation of multiplication,

$$24 \div 3 = b \quad \text{and} \quad b = 8 \quad \text{(see next page)}$$

Complete the problems below. Completed problems appear on page 58.

Sample Set

(a) $\dfrac{2}{5} \times \dfrac{3}{3} = \dfrac{\square}{\square}$

(b) $\dfrac{3}{4} \times \dfrac{\square}{4} = \dfrac{12}{\square}$

(c) $\dfrac{5}{8} \times \dfrac{\square}{\square} = \dfrac{30}{48}$

(d) $\dfrac{4}{11} \times \dfrac{\square}{\square} = \dfrac{\square}{55}$

(e) $\dfrac{6}{7} = \dfrac{\square}{42}$

(f) $\dfrac{13}{15} = \dfrac{\square}{180}$

The numerator can now be determined.

$$\frac{2}{3} \times \frac{8}{8} = \frac{16}{24}$$

What steps have taken place?

       new denominator old denominator

24 was divided by 3 to obtain 8

2 was multiplied by 8 to obtain 16

  old numerator   new numerator

This observation provides you with a shortcut. Try not to lose sight of the principles involved.

Sample Problem 2-7   Write $\frac{3}{8}$ as an equivalent fraction with denominator 144.

Solution:
$$144 \div 8 = 18$$
$$3 \times 18 = 54 \qquad \frac{3}{8} = \frac{54}{144}$$

Completed Problems

(a) $\dfrac{2}{5} \times \dfrac{3}{3} = \dfrac{6}{15}$

(b) $\dfrac{3}{4} \times \dfrac{4}{4} = \dfrac{12}{16}$

(c) $\dfrac{5}{8} \times \dfrac{6}{6} = \dfrac{30}{48}$

(d) $\dfrac{4}{11} \times \dfrac{5}{5} = \dfrac{20}{55}$

(e) $\dfrac{6}{7} = \dfrac{36}{42}$

(f) $\dfrac{13}{15} = \dfrac{156}{180}$

Name: _____

Class: _____

Exercises for Sec. 2-3 Fill in the missing numbers.

1. $\dfrac{2}{3} = \dfrac{}{15}$
2. $\dfrac{2}{3} = \dfrac{}{24}$
3. $\dfrac{2}{3} = \dfrac{}{27}$

4. $\dfrac{2}{3} = \dfrac{}{180}$
5. $\dfrac{3}{4} = \dfrac{}{20}$
6. $\dfrac{3}{4} = \dfrac{}{84}$

7. $\dfrac{3}{5} = \dfrac{}{25}$
8. $\dfrac{3}{5} = \dfrac{}{125}$
9. $\dfrac{3}{8} = \dfrac{}{72}$

10. $\dfrac{5}{8} = \dfrac{}{72}$
11. $\dfrac{2}{7} = \dfrac{}{42}$
12. $\dfrac{2}{9} = \dfrac{}{72}$

13. $\dfrac{5}{9} = \dfrac{}{180}$
14. $\dfrac{3}{10} = \dfrac{}{120}$
15. $\dfrac{7}{12} = \dfrac{}{60}$

16. $\dfrac{8}{11} = \dfrac{}{99}$
17. $\dfrac{5}{13} = \dfrac{}{91}$
18. $\dfrac{11}{20} = \dfrac{}{140}$

19. $\dfrac{3}{29} = \dfrac{}{87}$
20. $\dfrac{5}{41} = \dfrac{}{164}$
21. $\dfrac{17}{19} = \dfrac{}{152}$

22. $\dfrac{15}{23} = \dfrac{}{161}$ 23. $\dfrac{9}{47} = \dfrac{}{141}$ 24. $\dfrac{13}{51} = \dfrac{}{255}$

25. $\dfrac{15}{28} = \dfrac{}{252}$ 26. $\dfrac{21}{32} = \dfrac{}{288}$ 27. $\dfrac{31}{42} = \dfrac{}{168}$

28. $\dfrac{13}{36} = \dfrac{}{144}$ 29. $\dfrac{41}{45} = \dfrac{}{360}$ 30. $\dfrac{23}{48} = \dfrac{}{240}$

31. $\dfrac{17}{48} = \dfrac{}{288}$ 32. $\dfrac{19}{54} = \dfrac{}{432}$ 33. $\dfrac{17}{56} = \dfrac{}{504}$

34. $\dfrac{41}{60} = \dfrac{}{360}$ 35. $\dfrac{33}{64} = \dfrac{}{576}$ 36. $\dfrac{53}{72} = \dfrac{}{432}$

37. $\dfrac{31}{81} = \dfrac{}{648}$ 38. $\dfrac{13}{84} = \dfrac{}{504}$ 39. $\dfrac{23}{96} = \dfrac{}{480}$

40. $\dfrac{31}{100} = \dfrac{}{1200}$

Section 2-4 Least Common Multiples

A **common multiple** of two numbers is a number that is divisible by both. Some multiples of 6 and 9 are listed below. Their common multiples are in bold type.

Multiples of 6:   6,   12,   **18,**   24,   30,   **36,**   42,   48,   **54,**   60
Multiples of 9:   9,   **18,**   27,   **36,**   45,   **54,**   63

The **least common multiple** (LCM) of two numbers is the smallest number that is a multiple of both.

In the example above,

18 is the LCM of 6 and 9

> **The least common multiple of two numbers is the smallest number divisible by both.**

Observe the prime factorizations of 6, 9, and their least common multiple 18, with like prime factors grouped as shown below:

$$6 = (2) \mid (3)$$
$$9 = \mid (3)(3)$$
$$\text{LCM} = 18 = (2) \mid (3)(3)$$

Note that the factorization of the LCM has the greatest number of 2s that appears in either the factorization of 6 or the factorization of 9 (one 2 in 6). The factorization also has the greatest number of 3s in either factorization (two 3s in 9).

A knowledge of the multiplication tables is sufficient to find the LCM of small numbers. Can you find the LCM of

27   and   24

You could list multiples of each until the LCM appeared in each list. The technique demonstrated below is much simpler.

Sample Problem 2-8

Find the LCM of 24 and 27.

Solution:

Write the prime factorizations of 24 and 27 with like primes grouped as shown:

$$24 = (2)(2)(2) \mid (3)$$
$$27 = \mid (3)(3)(3)$$
$$\text{LCM} = (2)(2)(2) \mid (3)(3)(3)$$
$$= 216$$

Observe the quotients below:

$$216 \div 24 = \frac{216}{24} = \frac{\cancel{(2)}\cancel{(2)}\cancel{(2)}\cancel{(3)}(3)(3)}{\cancel{(2)}\cancel{(2)}\cancel{(2)}\cancel{(3)}} = 9$$

Prime Factorization, Equivalent Fractions, Least Common Multiple

$$216 \div 27 = \frac{216}{27} = \frac{(2)(2)(2)(\cancel{3})(\cancel{3})(\cancel{3})}{(\cancel{3})(\cancel{3})(\cancel{3})} = 8$$

If you had listed multiples, 216 would have been the ninth multiple of 24 listed, and the eighth multiple of 27.

Sample Problem 2-8

Solution:

Find the LCM of 60 and 18.

$$60 = (2)(2) \;\; (3) \quad\;\; (5)$$
$$18 = \;\;\;\;\; (2) \;\; (3)(3)$$
$$\text{LCM} = (2)(2) \;\; (3)(3) \;\; (5) = 180$$

Sample Problem 2-9

Solution:

Find the LCM of 36, 24, and 27.

$$36 = (2)(2) \qquad\;\; (3)(3)$$
$$24 = (2)(2)(2) \;\; (3)$$
$$27 = \qquad\qquad\;\; (3)(3)(3)$$
$$\text{LCM} = (2)(2)(2) \;\; (3)(3)(3) = 216$$

Sample Problem 2-10

Solution:

Find the LCM of 15 and 14.

$$15 = \qquad\;\; (3)(5)$$
$$14 = (2) \qquad\qquad\;\; (7)$$
$$\text{LCM} = (2) \;\; (3)(5) \;\; (7) = 210$$

When the two numbers have no common factors, the LCM is their product.

Sample Problem 2-11

Solution:

The number 7 is a prime, so it is simply listed:

Find the LCM of 7, 15, and 6.

$$7 = \qquad\qquad\;\;\; (7)$$
$$15 = \;\;\;\; (3)(5)$$
$$6 = (2)(3)$$
$$\text{LCM} = (2)(3)(5)(7) = 210$$

Review To find the LCM:

1. Write the prime factorization of each number.
2. The number of times a given prime will appear in the LCM is the greatest number of times the prime appears in **any one** of these factorizations.

Complete the factorizations below. Completed factorizations appear on page 64.

Sample Set

(a)   2 = (2)
        3 = ___( )___
        LCM = ( )( ) = ☐

(b)   4 = (2)( )
        6 = ___(2)( )___
        LCM = ( )( )( ) = ☐

(c)   12 = (2)(2)( )
        18 = ___(2)(3)( )___
        LCM = ( )( )( )( ) = ☐

(d)   36 =   ( )(2)( )(3)
        48 = ___( )( )( )(2)(3)___
        LCM = ( )( )( )( )( )( ) = ☐

(e)   12 = ( )( )(3)
        15 =   (3)   ( )
        18 = ___(2)( )( )___
        LCM = ( )( )( )( )( ) = ☐

(f)   24 = ( )( )(2)(3)
        30 =   (2)   ( )   ( )
        36 = ___( )( )(3)( )___
        LCM = ( )( )( )( )( )( ) = ☐

Completed Factorizations

(a) $2 = (2)$
        $3 = (3)$
       LCM $= (2)(3) = \boxed{6}$

(b) $4 = (2)(2)$
        $6 = (2)(3)$
       LCM $= (2)(2)(3) = \boxed{12}$

(c) $12 = (2)(2)(3)$
       $18 = (2)(3)(3)$
      LCM $= (2)(2)(3)(3) = \boxed{36}$

(d) $36 = (2)(2)(3)(3)$
       $48 = (2)(2)(2)(2)(3)$
     LCM $= (2)(2)(2)(2)(3)(3) = \boxed{144}$

(e) $12 = (2)(2)(3)$
       $15 = (3)(5)$
       $18 = (2)(3)(3)$
     LCM $= (2)(2)(3)(3)(5) = \boxed{180}$

(f) $24 = (2)(2)(2)(3)$
       $30 = (2)(3)(5)$
       $36 = (2)(2)(3)(3)$
    LCM $= (2)(2)(2)(3)(3)(5) = \boxed{360}$

Name: _____

Class: _____

Exercises for Sec. 2-4 Find the LCM of the following numbers:

1. 5, 9 2. 8, 9 3. 8, 12

4. 9, 12 5. 15, 18 6. 10, 14

7. 12, 16 8. 20, 25 9. 21, 16

10. 20, 24 11. 21, 28 12. 36, 48

13. 32, 48 14. 28, 49 15. 56, 48

16. 64, 56 17. 54, 64 18. 64, 72

19. 54, 72 20. 72, 84 21. 108, 144

22. 2, 3, 5 23. 4, 6, 10 24. 5, 10, 20

25. 8, 6, 9   26. 12, 15, 18   27. 12, 16, 18

28. 12, 20, 30   29. 18, 15, 20   30. 20, 15, 12

31. 24, 36, 30   32. 28, 49, 12   33. 28, 49, 42

34. 30, 24, 20   35. 16, 32, 64   36. 36, 48, 54

37. 32, 48, 64   38. 36, 54, 60   39. 54, 72, 96

40. 54, 81, 84

Name: _____

Class: _____

Unit 2 Test Write the prime factorization of each number.

1. 14

2. 20

3. 36

4. 81

5. 216

Find the LCM of the numbers.

6. 8, 12

7. 20, 28

8. 56, 70

9. 8, 12, 18

10. 24, 30, 36

Reduce each fraction to lowest terms.

11. $\dfrac{4}{12}$

12. $\dfrac{18}{24}$

13. $\dfrac{36}{20}$

14. $\dfrac{72}{54}$

15. $\dfrac{216}{288}$

68   Prime Factorization, Equivalent Fractions, Least Common Multiple

Fill in the missing numbers.

16. $\dfrac{4}{9} = \dfrac{}{36}$

17. $\dfrac{7}{8} = \dfrac{}{72}$

18. $\dfrac{19}{24} = \dfrac{}{120}$

19. $\dfrac{7}{54} = \dfrac{}{270}$

20. $\dfrac{11}{72} = \dfrac{}{504}$

Section 3-1 **Multiplication of Fractions**

The multiplication rule for fractions is given below.

Multiplication Rule
$$\frac{a}{b} \times \frac{c}{d} = \frac{a \times c}{b \times d}$$

Examples with common fractions are given below.

(a) $\dfrac{2}{3} \times \dfrac{5}{7} = \dfrac{2 \times 5}{3 \times 7} = \dfrac{10}{21}$

(b) $\dfrac{5}{9} \times \dfrac{13}{7} = \dfrac{65}{63}$

When multiplying a fraction by a whole number, write the whole number over 1.

Examples with whole numbers and common fractions are given below:

(a) $5 \times \dfrac{3}{4} = \dfrac{5}{1} \times \dfrac{3}{4} = \dfrac{15}{4}$

(b) $\dfrac{7}{8} \times 3 = \dfrac{7}{8} \times \dfrac{3}{1} = \dfrac{21}{8}$

(c) $\dfrac{5}{9} \times 7 \times \dfrac{2}{11} = \dfrac{5}{9} \times \dfrac{7}{1} \times \dfrac{2}{11} = \dfrac{70}{99}$

An improper fraction is an acceptable representation of a number. You may write improper fractions as mixed numbers if you wish. Mixed numbers will be discussed in the next unit.

Interpretation A single fraction has been interpreted as representing a part of one whole object.

$$\tfrac{1}{2} = \tfrac{1}{2} \times 1 = \tfrac{1}{2} \text{ of 1 object}$$

How might the product of two fractions be interpreted? Consider

$$\tfrac{3}{5} \times \tfrac{1}{2} = \tfrac{3}{10}$$

In the diagram below, the shaded area represents $\tfrac{3}{5}$ of $\tfrac{1}{2}$ of a whole object:

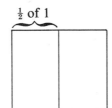
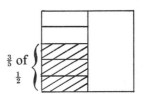

Observe that $\tfrac{3}{5}$ of $\tfrac{1}{2}$ of the rectangle is the same as $\tfrac{3}{10}$ of the rectangle.

72 Arithmetic of Common Fractions

A "fractional part of a fraction" can be found by multiplying the fractions.

$$\tfrac{3}{5} \text{ of } \tfrac{1}{2} = \tfrac{3}{5} \times \tfrac{1}{2} = \tfrac{3}{10}$$

In the word problems that will come later, the word *of* most often indicates that multiplication operation is to be performed.

Cancellation May Be Used

Cancellation "up and down" is equivalent to reducing a fraction before multiplying.

$$(a) \quad \frac{2}{7} \times \frac{\cancel{3}^{1}}{\cancel{15}_{5}} = \frac{2 \times 1}{7 \times 5} = \frac{2}{35}$$

$$(b) \quad \frac{5}{11} \times \frac{\cancel{\cancel{36}}^{\,3}_{\,\cancel{18}}}{\cancel{\cancel{84}}_{\,\cancel{42}}^{\,7}} = \frac{15}{77}$$

A factor in the numerator of one fraction may be canceled with a like factor in the denominator of another.

$$\frac{\cancel{3}^{1}}{5} \times \frac{7}{\cancel{3}_{1}} = \frac{1 \times 7}{5 \times 1} = \frac{7}{5}$$

This can be justified as follows:

$$\frac{3}{5} \times \frac{7}{3} = \frac{(3)(7)}{(5)(3)} \qquad \textbf{multiplication rule}$$

$$= \frac{(7)(3)}{(5)(3)} \qquad \textbf{commutative law of multiplication}$$

$$= \frac{7}{5} \times \frac{3}{3} \qquad \textbf{multiplication rule}$$

$$= \frac{7}{5} \times 1$$

$$= \frac{7}{5}$$

As in the case of reducing a fraction, this cancellation is a shortcut dependent on the fact that $\tfrac{3}{3} = 1$.

Sec. 3-1 Multiplication of Fractions

Sample Problem 3-1 Multiply $\frac{8}{9} \times \frac{15}{16}$.

Solution: Cancel the common factor 8 (of 8 and 16):
Cancel the common factor 3 (of 9 and 15):

$$\frac{\overset{1}{\cancel{8}}}{\underset{3}{\cancel{9}}} \times \frac{\overset{5}{\cancel{15}}}{\underset{2}{\cancel{16}}} = \frac{5}{6}$$

See sample problems 3-2 and 3-3 (pp. 74 and 75) before completing the sample set below.

Complete the problems below. The completed problems are on page 75.

Sample Set

Step 1. Cancel 4s:
$$\frac{18}{\cancel{20}} \times \frac{54}{36} \times \frac{\overset{7}{\cancel{28}}}{63}$$
$\square$

Step 2. Cancel 7s:
$$\frac{18}{\cancel{20}} \times \frac{54}{36} \times \frac{\overset{\overset{1}{\cancel{7}}}{\cancel{28}}}{\cancel{63}}$$
$\square \qquad \square$

Step 3. Cancel 18s:
$$\frac{\overset{1}{\cancel{18}}}{\cancel{20}} \times \frac{54}{\cancel{36}} \times \frac{\overset{\overset{1}{\cancel{7}}}{\cancel{28}}}{\cancel{63}}$$
$\square \qquad \square \qquad \square$

Step 4. Cancel 9s:
$$\frac{\overset{1}{\cancel{18}}}{\cancel{20}} \times \frac{\overset{\square}{\cancel{54}}}{\cancel{36}} \times \frac{\overset{\overset{1}{\cancel{7}}}{\cancel{28}}}{\underset{1}{\cancel{63}}}$$
$\square \qquad \square$

Step 5. Cancel 2s:
$$\frac{\overset{1}{\cancel{18}}}{\cancel{20}} \times \frac{\overset{\overset{\square}{\cancel{}}}{\cancel{54}}}{\cancel{36}} \times \frac{\overset{\overset{1}{\cancel{7}}}{\cancel{28}}}{\underset{1}{\cancel{63}}} = \frac{\square}{\square}$$
$\square \qquad \underset{1}{\cancel{}} \qquad \underset{1}{\cancel{}}$

74 Arithmetic of Common Fractions

Sample Problem 3-2

Multiply $\frac{30}{48} \times \frac{28}{20}$.

Solution:

Step 1. Cancel the common factor 10 (of 30 and 20):
Cancel the common factor 4 (of 48 and 28):

$$\frac{\overset{3}{\cancel{30}}}{\underset{12}{\cancel{48}}} \times \frac{\overset{7}{\cancel{28}}}{\underset{2}{\cancel{20}}}$$

Step 2. Cancel the common factor 3 (of 3 and 12):

$$\frac{\overset{\overset{1}{\cancel{3}}}{\cancel{30}}}{\underset{\underset{4}{\cancel{12}}}{\cancel{48}}} \times \frac{\overset{7}{\cancel{28}}}{\underset{2}{\cancel{20}}} = \frac{7}{8}$$

In step 1 of the above problem, could the choice of factors to be canceled have been different? The answer is yes.

$$\frac{\overset{5}{\cancel{30}}}{\underset{8}{\cancel{48}}} \times \frac{\overset{7}{\cancel{28}}}{\underset{5}{\cancel{20}}} \quad \begin{array}{l}\text{cancel 6 (in 30 and 48)}\\ \text{cancel 4 (in 28 and 20)}\end{array}$$

$$= \frac{\overset{\overset{1}{\cancel{5}}}{\cancel{30}}}{\underset{8}{\cancel{48}}} \times \frac{7}{\underset{\underset{1}{\cancel{5}}}{\cancel{20}}} = \frac{7}{8} \quad \text{cancel 5}$$

If factors are canceled correctly, the final result will be the same.

Warning

Do not cancel a factor in the numerator of one fraction with a like factor in the numerator of another fraction. The same holds for denominators.

The final product should be reduced to lowest terms.

If all possible cancellations are carried out before multiplying, the product will be in lowest terms.

Sec. 3-1 Multiplication of Fractions

Sample Problem 3-3

Possible Steps:

Multiply $\frac{12}{15} \times \frac{9}{36} \times \frac{20}{16}$.

Step 1. Cancel 4 (in 12 and 16):
Cancel 5 (in 15 and 20):
Cancel 9 (in 9 and 36):

$$\frac{\cancel{12}^{3}}{\cancel{15}_{3}} \times \frac{\cancel{9}^{1}}{\cancel{36}_{4}} \times \frac{\cancel{20}^{4}}{\cancel{16}_{4}}$$

Step 2. Cancel 3:
Cancel the 4 in the numerator with **one** 4 in a denominator:

$$\frac{\cancel{\cancel{12}}^{1}}{\cancel{\cancel{15}}_{1}} \times \frac{\cancel{9}^{1}}{\cancel{36}_{4}} \times \frac{\cancel{\cancel{20}}^{1}}{\cancel{\cancel{16}}_{4}} = \frac{1}{4}$$

Completed Problems

Step 1. Cancel 4s:
$$\frac{18}{\cancel{20}_{5}} \times \frac{54}{36} \times \frac{\cancel{28}^{7}}{63}$$

Step 2. Cancel 7s:
$$\frac{18}{\cancel{20}_{5}} \times \frac{54}{36} \times \frac{\cancel{\cancel{28}}^{1}}{\cancel{63}_{9}}$$

Step 3. Cancel 18s:
$$\frac{\cancel{18}^{1}}{\cancel{20}_{5}} \times \frac{54}{\cancel{36}_{2}} \times \frac{\cancel{\cancel{28}}^{1}}{\cancel{63}_{9}}$$

Step 4. Cancel 9s:
$$\frac{\cancel{18}^{1}}{\cancel{20}_{5}} \times \frac{\cancel{54}^{6}}{\cancel{36}_{2}} \times \frac{\cancel{\cancel{28}}^{1}}{\cancel{\cancel{63}}_{1}}$$

Step 5. Cancel 2s:
$$\frac{\cancel{18}^{1}}{\cancel{20}_{5}} \times \frac{\cancel{\cancel{54}}^{3}}{\cancel{\cancel{36}}_{1}} \times \frac{\cancel{\cancel{28}}^{1}}{\cancel{\cancel{63}}_{1}} = \frac{3}{5}$$

Name: _____

Class: _____

Exercises for Sec. 3-1 Find the products. Express each answer in lowest terms.

1. $\dfrac{3}{4} \times \dfrac{4}{6} =$

2. $\dfrac{6}{8} \times \dfrac{4}{9} =$

3. $\dfrac{2}{5} \times \dfrac{3}{7} =$

4. $\dfrac{6}{9} \times 3 =$

5. $\dfrac{7}{8} \times \dfrac{4}{5} =$

6. $\dfrac{4}{10} \times \dfrac{5}{6} =$

7. $\dfrac{7}{8} \times \dfrac{5}{7} =$

8. $\dfrac{9}{8} \times \dfrac{4}{6} =$

9. $8 \times \dfrac{5}{20} =$

10. $\dfrac{8}{12} \times \dfrac{9}{16} =$

11. $\dfrac{3}{9} \times \dfrac{12}{16} =$

12. $\dfrac{12}{9} \times 6 =$

13. $\dfrac{18}{20} \times \dfrac{15}{36} =$

14. $\dfrac{20}{21} \times \dfrac{28}{16} =$

15. $\dfrac{15}{30} \times \dfrac{12}{18} =$

16. $\dfrac{28}{45} \times \dfrac{15}{42} =$

17. $6 \times \dfrac{48}{54} =$

18. $\dfrac{60}{48} \times \dfrac{27}{36} =$

19. $\dfrac{63}{54} \times \dfrac{48}{56} =$

20. $\dfrac{36}{81} \times \dfrac{63}{32} =$

21. $\dfrac{2}{3} \times \dfrac{4}{5} \times \dfrac{5}{8} =$

22. $\dfrac{1}{3} \times \dfrac{2}{4} \times \dfrac{3}{5} =$

23. $\dfrac{1}{5} \times 15 \times \dfrac{1}{3} =$

24. $\dfrac{2}{6} \times \dfrac{3}{7} \times \dfrac{14}{8} =$

25. $\dfrac{3}{9} \times \dfrac{6}{10} \times 20 =$

26. $\dfrac{3}{4} \times \dfrac{6}{10} \times \dfrac{5}{9} =$

27. $\dfrac{9}{10} \times \dfrac{4}{6} \times \dfrac{5}{8} =$

28. $\dfrac{9}{12} \times \dfrac{1}{18} \times \dfrac{16}{10} =$

29. $\dfrac{12}{15} \times \dfrac{5}{18} \times \dfrac{9}{20} =$

30. $\dfrac{15}{16} \times \dfrac{12}{18} \times \dfrac{9}{10} =$

31. $\dfrac{10}{18} \times \dfrac{21}{15} \times \dfrac{12}{14} =$

32. $\dfrac{15}{20} \times \dfrac{14}{21} \times \dfrac{8}{9} =$

33. $\dfrac{16}{20} \times \dfrac{5}{18} \times \dfrac{1}{8} =$

34. $\dfrac{12}{20} \times \dfrac{15}{16} \times \dfrac{4}{18} =$

35. $\dfrac{16}{12} \times \dfrac{15}{18} \times \dfrac{5}{20} =$

36. $\dfrac{18}{21} \times \dfrac{14}{20} \times \dfrac{5}{6} =$

37. $\dfrac{15}{20} \times \dfrac{16}{18} \times \dfrac{21}{28} =$

38. $\dfrac{18}{24} \times \dfrac{32}{36} \times \dfrac{28}{49} =$

39. $\dfrac{42}{48} \times \dfrac{20}{36} \times \dfrac{32}{35} =$

40. $\dfrac{24}{36} \times \dfrac{63}{64} \times \dfrac{48}{56} =$

Section 3-2 Division of Fractions

The division rule for fractions is given below.

Division Rule
$$\frac{a}{b} \div \frac{c}{d} = \frac{a}{b} \times \frac{d}{c} = \frac{a \times d}{b \times c}$$

This rule will be justified at the end of the section.

The fraction d/c is the **reciprocal** of c/d.

Sample Problem 3-4

Divide $\frac{24}{15} \div \frac{16}{20}$.

Solution: Multiply $\frac{24}{15}$ by the reciprocal of $\frac{16}{20}$:

$$\frac{24}{15} \div \frac{16}{20} = \frac{\cancel{24}}{\cancel{15}} \times \frac{\cancel{20}}{\cancel{16}} = 2$$

Sample Problem 3-5

Divide $\frac{8}{5} \div 4$.

Solution: Multiply $\frac{8}{5}$ by the reciprocal of 4:

$$\frac{8}{5} \div \frac{4}{1} = \frac{\cancel{8}}{5} \times \frac{1}{\cancel{4}} = \frac{2}{5}$$

Recall the definition of division given in Unit 2:

$$A \div B = \frac{A}{B}$$

This definition is not restricted to the case in which A and B are whole numbers. A and B can be fractions. For example,

$$\frac{2}{3} \div \frac{7}{5} = \frac{\frac{2}{3}}{\frac{7}{5}} \quad \textbf{a number}$$

A fraction in which the numerator or denominator (or both) are not whole numbers is called a **complex fraction**.

Since the "value" of a complex fraction is not easily seen, it is convenient to express such fractions as equivalent common fractions. Recall that the numerators and denominators of common fractions are whole numbers.

82 Arithmetic of Common Fractions

Sample Problem 3-6 Write $\frac{8/9}{7/11}$ as an equivalent common fraction.

Solution:
$$\frac{\frac{8}{9}}{\frac{7}{11}} = \frac{8}{9} \div \frac{7}{11} = \frac{8}{9} \times \frac{11}{7} = \frac{88}{63}$$

Complete the problems below. Completed problems are on page 84.

Sample Set

(a) $\dfrac{3}{5} \div \dfrac{2}{7} = \dfrac{3}{5} \times \dfrac{\square}{\square} = \dfrac{\square}{\square}$

(b) $\dfrac{7}{9} \div \dfrac{5}{3} = \dfrac{\square}{\square} \times \dfrac{\cancel{3}^{1}}{5} = \dfrac{\square}{\square}$

(c) $\dfrac{5}{8} \div \dfrac{7}{12} = \dfrac{5}{8} \times \dfrac{\cancel{\square}}{\square} = \dfrac{\square}{\square}$

(d) $\dfrac{7}{12} \div \dfrac{7}{16} = \dfrac{\cancel{7}^{1}}{\cancel{12}_{3}} \times \dfrac{\cancel{\square}}{\square} = \dfrac{\square}{\square}$

Justification of the Division Rule

It was stated earlier that the fraction $\dfrac{B}{B}$ is equivalent to 1. For example,

$$\frac{B}{B} = 1$$

Here, B is not restricted to being a whole number. For example,

$$\frac{\frac{2}{3}}{\frac{2}{3}} = \frac{2}{3} \div \frac{2}{3} = \frac{\cancel{2}^{1}}{\cancel{3}_{1}} \times \frac{\cancel{3}^{1}}{\cancel{2}_{1}} = 1$$

The justification for the division rule can now be given.

$$\frac{a}{b} \div \frac{c}{d} = \frac{\frac{a}{b}}{\frac{c}{d}} \qquad \textbf{definition of division}$$

$$= \frac{\frac{a}{b}}{\frac{c}{d}} \times \frac{\frac{d}{c}}{\frac{d}{c}} \qquad \frac{d/c}{d/c} = 1$$

$$= \frac{\dfrac{a}{b} \times \dfrac{d}{c}}{\dfrac{c}{d} \times \dfrac{d}{c}} \qquad \text{\textbf{multiplication rule for all fractions}}$$

$$= \frac{\dfrac{a}{b} \times \dfrac{d}{c}}{1} \qquad \frac{c}{d} \times \frac{d}{c} = \frac{c \times d}{d \times c} = 1$$

$$= \frac{a}{b} \times \frac{d}{c}$$

Notice that a key step in the demonstration above depended on the use of the multiplication rule for fractions. You may well ask if the multiplication rule can be justified, or if it is a law. It can be justified in terms of simple properties. This would be quite complicated, and therefore will not be done.

Completed Problems

(a) $\dfrac{3}{5} \div \dfrac{2}{7} = \dfrac{3}{5} \times \dfrac{7}{2} = \dfrac{21}{10}$

(b) $\dfrac{7}{9} \div \dfrac{5}{3} = \dfrac{7}{\underset{3}{9}} \times \dfrac{\overset{1}{3}}{5} = \dfrac{7}{15}$

(c) $\dfrac{5}{8} \div \dfrac{7}{12} = \dfrac{5}{\underset{2}{8}} \times \dfrac{\overset{3}{12}}{7} = \dfrac{15}{14}$

(d) $\dfrac{7}{12} \div \dfrac{7}{16} = \dfrac{\overset{1}{7}}{\underset{3}{12}} \times \dfrac{\overset{4}{16}}{\underset{1}{7}} = \dfrac{4}{3}$

Name: _____

Class: _____

Exercises for Sec. 3-2 Find the quotients. Express each answer in lowest terms.

1. $\dfrac{1}{2} \div \dfrac{1}{3} =$

2. $\dfrac{2}{7} \div \dfrac{3}{5} =$

3. $\dfrac{11}{9} \div \dfrac{3}{5} =$

4. $\dfrac{3}{6} \div \dfrac{3}{4} =$

5. $\dfrac{2}{5} \div \dfrac{8}{10} =$

6. $\dfrac{4}{9} \div \dfrac{5}{6} =$

7. $2 \div \dfrac{2}{5} =$

8. $\dfrac{7}{12} \div \dfrac{5}{8} =$

9. $\dfrac{8}{3} \div 4 =$

10. $\dfrac{9}{12} \div \dfrac{2}{3} =$

11. $\dfrac{3}{12} \div \dfrac{5}{8} =$

12. $\dfrac{7}{9} \div 3 =$

13. $\dfrac{8}{10} \div \dfrac{6}{5} =$

14. $\dfrac{12}{16} \div \dfrac{9}{10} =$

15. $\dfrac{8}{15} \div \dfrac{4}{10} =$

16. $\dfrac{15}{18} \div \dfrac{10}{24} =$

17. $\dfrac{15}{18} \div \dfrac{12}{9} =$

18. $\dfrac{12}{20} \div \dfrac{16}{15} =$

19. $\dfrac{15}{27} \div \dfrac{20}{18} =$

20. $\dfrac{25}{15} \div 20 =$

21. $36 \div \dfrac{27}{12} =$

22. $\dfrac{21}{28} \div \dfrac{9}{12} =$

23. $\dfrac{21}{28} \div \dfrac{10}{18} =$

24. $\dfrac{32}{56} \div \dfrac{20}{28} =$

25. $\dfrac{54}{36} \div 9 =$

26. $81 \div \dfrac{72}{64} =$

27. $\dfrac{54}{72} \div \dfrac{36}{48} =$

28. $\dfrac{56}{72} \div \dfrac{28}{27} =$

29. $\dfrac{56}{48} \div \dfrac{21}{36} =$

30. $\dfrac{72}{81} \div \dfrac{20}{27} =$

Write each complex fraction as an equivalent common fraction.

31. $\dfrac{\frac{1}{2}}{\frac{1}{3}} =$

32. $\dfrac{\frac{5}{6}}{\frac{10}{9}} =$

33. $\dfrac{\frac{2}{3}}{2} =$

34. $\dfrac{\frac{2}{5}}{\frac{7}{3}} =$

35. $\dfrac{\frac{4}{6}}{\frac{8}{9}} =$

36. $\dfrac{\frac{1}{3}}{\frac{10}{9}} =$

37. $\dfrac{\frac{8}{9}}{\frac{6}{7}} =$

38. $\dfrac{\frac{3}{4}}{\frac{12}{15}} =$

39. $\dfrac{\frac{5}{15}}{7} =$

40. $\dfrac{\frac{3}{4}}{\frac{6}{4}} =$

41. $\dfrac{\frac{5}{6}}{\frac{15}{12}} =$

42. $\dfrac{\frac{12}{8}}{3} =$

43. $\dfrac{\frac{12}{16}}{\frac{9}{24}} =$

44. $\dfrac{\frac{16}{18}}{\frac{24}{27}} =$

45. $\dfrac{\frac{15}{18}}{\frac{20}{12}} =$

46. $\dfrac{\frac{20}{21}}{\frac{8}{14}} =$

47. $\dfrac{\frac{32}{36}}{\frac{24}{18}} =$

48. $\dfrac{\frac{18}{24}}{\frac{20}{30}} =$

Section 3-3 Addition of Fractions

The denominator of a fraction can be interpreted as the number of equal parts into which a whole unit is divided.

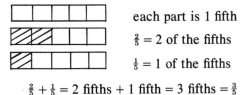

each part is 1 fifth
$\frac{2}{5}$ = 2 of the fifths
$\frac{1}{5}$ = 1 of the fifths

$\frac{2}{5} + \frac{1}{5}$ = 2 fifths + 1 fifth = 3 fifths = $\frac{3}{5}$

When the denominator of two or more fractions is the same, the denominator is called a **common denominator.**

The following addition rule holds for two fractions with a common denominator.

Addition Rule
$$\frac{a}{c} + \frac{b}{c} = \frac{a+b}{c}$$

This rule will be justified at the end of this section.

Examples

(a) $\dfrac{1}{8} + \dfrac{4}{8} = \dfrac{1+4}{8} = \dfrac{5}{8}$

(b) $\dfrac{7}{16} + \dfrac{5}{16} = \dfrac{12}{16} = \dfrac{3}{4}$

(c) $\dfrac{9}{5} + \dfrac{3}{5} = \dfrac{12}{5}$

Rules similar to the one stated above could be shown to hold when any number of fractions is added.

Examples

(a) $\frac{2}{9} + \frac{1}{9} + \frac{4}{9} = \frac{7}{9}$

(b) $\frac{7}{16} + \frac{3}{16} + \frac{1}{16} + \frac{3}{16} = \frac{14}{16} = \frac{7}{8}$

(c) $\frac{9}{5} + \frac{7}{5} + \frac{2}{5} = \frac{18}{5}$

The parts of 1 represented by the fractions below are not the same.

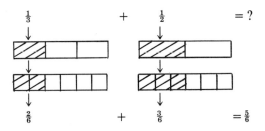

Each fraction was written as an equivalent fraction with common denominator 6. The addition rule was then applied. Other common denominators could have been used. For example,

$$\frac{4}{12} + \frac{6}{12} = \frac{10}{12} = \frac{5}{6}$$

$$\frac{6}{18} + \frac{9}{18} = \frac{15}{18} = \frac{5}{6}$$

The smallest common denominator for two or more fractions is referred to as the **least common denominator** (LCD).

> **LCD = LCM**
>
> The least common denominator
> for a set of denominators is
> the least common multiple
> of those denominators.

Sample Problem 3-7 Find $\frac{3}{7} + \frac{2}{3}$.

Solution:

Step 1. The LCM of 7 and 3 is 21; write each fraction as an equivalent fraction with LCD = 21:

$$\frac{3}{7} \times \frac{3}{3} = \frac{9}{21}$$

$$\frac{2}{3} \times \frac{7}{7} = \frac{14}{21}$$

Step 2. Add $\frac{9}{21}$ and $\frac{14}{21}$:

$$\frac{9}{21} + \frac{14}{21} = \frac{23}{21}$$

The entire solution can be written on one line:

$$\frac{3}{7} + \frac{2}{3} = \underbrace{\frac{\overbrace{3(3)}^{9}}{\underbrace{7(3)}_{21}}}_{} + \underbrace{\frac{\overbrace{2(7)}^{14}}{\underbrace{3(7)}_{21}}}_{} = \frac{23}{21}$$

Do not add the denominators.

Sample Problem 3-8 Add $\frac{5}{6} + \frac{8}{9}$.

Solution:

Step 1. Find the LCM of 6 and 9:

$$6 = (2)(3)$$
$$9 = \underline{(3)(3)}$$
$$LCD = LCM = (2)(3)(3) = 18$$

you can probably do this step in your head

Step 2. Divide: $18 \div 6 = 3$ and $18 \div 9 = 2$

Step 3. Add $\dfrac{5}{6} + \dfrac{8}{9} = \dfrac{\overbrace{5(3)}^{15}}{\underbrace{6(3)}_{18}} + \dfrac{\overbrace{8(2)}^{16}}{\underbrace{9(2)}_{18}} = \dfrac{31}{18}$

Sec. 3-3 Addition of Fractions

Sample Problem 3-9

Find $\frac{5}{84} + \frac{7}{30}$.

Solution:

Step 1. Find the LCM of 84 and 30:
$$84 = (2)(2)(3)(7)$$
$$30 = (2)(3)(5)$$
$$\text{LCD} = \text{LCM} = (2)(2)(3)(5)(7) = 420$$

Step 2. Divide: $420 \div 84 = 5$ and $420 \div 30 = 14$

Step 3. Add $\dfrac{5}{84} + \dfrac{7}{30} = \dfrac{5(5)}{84(5)} + \dfrac{7(14)}{30(14)} = \dfrac{123}{420} = \dfrac{41}{140}$

(with $5(5)=25$, $84(5)=420$, $7(14)=98$, $30(14)=420$)

In Sample Problem 3-9, could you perform the division in step 2 in your head? This division can be done using cancellation.

$$420 \div 84 = \frac{420}{84} = \frac{(2)(2)(3)(5)(7)}{(2)(2)(3)(7)} = 5$$

$$420 \div 30 = \frac{420}{30} = \frac{(2)(2)(3)(5)(7)}{(2)(3)(5)} = 14$$

You probably do not have to write what appears above. Can you see the answers 5 and 14 by looking at the factorizations in step 1?

See sample problem 3-10 (p. 92) before completing the sample set below.

Complete the problems below. Completed problems appear on page 93.

Sample Set

(a) $\dfrac{2}{3} + \dfrac{4}{3} = \dfrac{\Box}{\Box} = \Box$

(b) $\dfrac{2}{7} + \dfrac{3}{7} = \dfrac{\Box}{\Box}$

(c) $\dfrac{3}{5} + \dfrac{1}{4} = \dfrac{\Box}{20} + \dfrac{\Box}{20} = \dfrac{\Box}{\Box}$

(d) $\dfrac{7}{12} + \dfrac{5}{18} = \dfrac{\Box}{\Box} + \dfrac{\Box}{\Box} = \dfrac{\Box}{\Box}$

(e) $\dfrac{5}{36} + \dfrac{3}{54} = \dfrac{\Box}{\Box} + \dfrac{\Box}{\Box} = \dfrac{\Box}{\Box} = \dfrac{\Box}{\Box}$

Sample Problem 3-10

Solution:

Find $\frac{5}{12} + \frac{4}{15} + \frac{3}{20}$.

Step 1. Find the LCM of 12, 15, and 20:
$$12 = (2)(2)(3)$$
$$15 = (3)(5)$$
$$20 = (2)(2)(5)$$
$$\text{LCD} = \text{LCM} = \overline{(2)(2)(3)(5)} = 60$$

Step 2. Divide: $60 \div 12 = 5$, $60 \div 15 = 4$, and $60 \div 20 = 3$

Step 3. Add and reduce:
$$\frac{5}{12} + \frac{4}{15} + \frac{3}{20} = \frac{\overbrace{5(5)}^{25}}{12(5)} + \frac{\overbrace{4(4)}^{16}}{15(4)} + \frac{\overbrace{3(3)}^{9}}{20(3)} = \frac{50}{60} = \frac{5}{6}$$

Justification of the Addition Rule

$$\frac{a}{c} + \frac{b}{c} = \frac{a+b}{c}$$

When a whole object is divided into c equal parts, the object may be thought of as being divided into units of size $1/c$. Then

$$\frac{a}{c} \text{ represents } a \text{ of the units } \frac{1}{c}$$

$$\frac{b}{c} \text{ represents } b \text{ of the units } \frac{1}{c}$$

$$\frac{a+b}{c} \text{ represents } a+b \text{ of the units } \frac{1}{c}$$

We then have

$$\frac{a}{c} + \frac{b}{c} = a \text{ of the units } \frac{1}{c} + b \text{ of the units } \frac{1}{c}$$

$$= a + b \text{ of the units } \frac{1}{c}$$

$$= \frac{a+b}{c}$$

A similar justification can be given using the definitions below.

$$\frac{a}{c} = a \times \frac{1}{c} \qquad \frac{b}{c} = b \times \frac{1}{c} \qquad \frac{a+b}{c} = (a+b) \times \frac{1}{c}$$

The line in the last fraction plays the same role as parentheses. It groups the addition in the numerator.

With these definitions in mind, the addition rule can be shown easily using the distributive law.

Sec. 3-3 Addition of Fractions

$$\frac{a}{c} + \frac{b}{c} = a \times \frac{1}{c} + b \times \frac{1}{c} \quad \textbf{definition}$$

$$= (a + b) \times \frac{1}{c} \quad \textbf{distributive law}$$

$$= \frac{a+b}{c}$$

Completed Problems

(a) $\quad \dfrac{2}{3} + \dfrac{4}{3} = \dfrac{\boxed{6}}{\boxed{3}} = \boxed{2}$ (b) $\quad \dfrac{2}{7} + \dfrac{3}{7} = \dfrac{\boxed{5}}{\boxed{7}}$

(c) $\quad \dfrac{3}{5} + \dfrac{1}{4} = \dfrac{\boxed{12}}{20} + \dfrac{\boxed{5}}{20} = \dfrac{\boxed{17}}{\boxed{20}}$

(d) $\quad \dfrac{7}{12} + \dfrac{5}{18} = \dfrac{\boxed{21}}{\boxed{36}} + \dfrac{\boxed{10}}{\boxed{36}} = \dfrac{\boxed{31}}{\boxed{36}}$

(e) $\quad \dfrac{5}{36} + \dfrac{3}{54} = \dfrac{\boxed{15}}{\boxed{108}} + \dfrac{\boxed{6}}{\boxed{108}} = \dfrac{\boxed{21}}{\boxed{108}} = \dfrac{\boxed{7}}{\boxed{36}}$

Name: _____

Class: _____

Exercises for Sec. 3-3 Find the sums. Express each answer in lowest terms.

1. $\dfrac{2}{9} + \dfrac{5}{9} =$

2. $\dfrac{7}{11} + \dfrac{3}{11} =$

3. $\dfrac{3}{8} + \dfrac{5}{8} =$

4. $\dfrac{3}{4} + \dfrac{1}{2} =$

5. $\dfrac{2}{3} + \dfrac{5}{6} =$

6. $\dfrac{2}{3} + \dfrac{1}{2} =$

7. $\dfrac{3}{9} + \dfrac{1}{3} =$

8. $\dfrac{2}{3} + \dfrac{3}{5} =$

9. $\dfrac{3}{4} + \dfrac{5}{6} =$ 10. $\dfrac{3}{8} + \dfrac{5}{6} =$

11. $\dfrac{1}{6} + \dfrac{5}{9} =$ 12. $\dfrac{4}{7} + \dfrac{9}{14} =$

13. $\dfrac{4}{9} + \dfrac{5}{12} =$ 14. $\dfrac{3}{8} + \dfrac{7}{12} =$

15. $\dfrac{4}{15} + \dfrac{3}{10} =$ 16. $\dfrac{1}{15} + \dfrac{7}{12} =$

Sec. 3-3 Addition of Fractions

17. $\dfrac{5}{12}+\dfrac{5}{18}=$

18. $\dfrac{9}{15}+\dfrac{5}{18}=$

19. $\dfrac{5}{12}+\dfrac{7}{15}=$

20. $\dfrac{7}{24}+\dfrac{5}{21}=$

21. $\dfrac{9}{25}+\dfrac{7}{30}=$

22. $\dfrac{11}{24}+\dfrac{5}{30}=$

23. $\dfrac{7}{24}+\dfrac{3}{18}=$

24. $\dfrac{13}{27}+\dfrac{5}{36}=$

25. $\dfrac{3}{20}+\dfrac{5}{28}=$

26. $\dfrac{7}{32}+\dfrac{5}{24}=$

27. $\dfrac{14}{27}+\dfrac{5}{36}=$

28. $\dfrac{11}{36}+\dfrac{5}{42}=$

29. $\dfrac{19}{72}+\dfrac{13}{54}=$

30. $\dfrac{11}{84}+\dfrac{7}{96}=$

31. $\dfrac{2}{3}+\dfrac{3}{4}+\dfrac{5}{6}=$

32. $\dfrac{1}{2}+\dfrac{1}{3}+\dfrac{1}{6}=$

33. $\dfrac{1}{2} + \dfrac{3}{4} + \dfrac{3}{8} =$

34. $\dfrac{3}{5} + \dfrac{2}{3} + \dfrac{4}{15} =$

35. $\dfrac{4}{9} + \dfrac{5}{8} + \dfrac{7}{12} =$

36. $\dfrac{5}{6} + \dfrac{3}{4} + \dfrac{1}{2} =$

37. $\dfrac{5}{12} + \dfrac{4}{15} + \dfrac{3}{20} =$

38. $\dfrac{7}{12} + \dfrac{5}{16} + \dfrac{7}{18} =$

39. $\dfrac{7}{12} + \dfrac{3}{21} + \dfrac{5}{28} =$

40. $\dfrac{11}{12} + \dfrac{7}{20} + \dfrac{2}{15} =$

41. $\dfrac{5}{16} + \dfrac{3}{20} + \dfrac{2}{15} =$

42. $\dfrac{4}{21} + \dfrac{5}{28} + \dfrac{7}{16} =$

43. $\dfrac{7}{36} + \dfrac{5}{27} + \dfrac{5}{72} =$

44. $\dfrac{9}{48} + \dfrac{7}{36} + \dfrac{1}{27} =$

45. $\dfrac{5}{72} + \dfrac{2}{81} + \dfrac{7}{54} =$

46. $\dfrac{1}{30} + \dfrac{3}{25} + \dfrac{5}{24} =$

47. $\dfrac{3}{4} + \dfrac{1}{2} + \dfrac{3}{8} + \dfrac{1}{16} =$

48. $\dfrac{3}{4} + \dfrac{5}{8} + \dfrac{7}{12} + \dfrac{1}{3} =$

Sec. 3-3 Additions of Fractions

49. $\dfrac{2}{3} + \dfrac{5}{9} + \dfrac{1}{6} + \dfrac{7}{18} =$

50. $\dfrac{1}{5} + \dfrac{3}{10} + \dfrac{2}{15} + \dfrac{2}{3} =$

51. $\dfrac{5}{8} + \dfrac{6}{9} + \dfrac{1}{12} + \dfrac{7}{18} =$

52. $\dfrac{9}{16} + \dfrac{5}{12} + \dfrac{7}{18} + \dfrac{11}{27} =$

53. $\dfrac{1}{12} + \dfrac{5}{24} + \dfrac{11}{16} + \dfrac{5}{18} =$

54. $\dfrac{3}{20} + \dfrac{5}{24} + \dfrac{7}{30} + \dfrac{1}{18} =$

55. $\dfrac{5}{36} + \dfrac{7}{48} + \dfrac{4}{27} + \dfrac{5}{72} =$

Section 3-4 Subtraction of Fractions

The subtraction rule for fractions with common denominator is given below.

Subtraction Rule

$$\frac{a}{c} - \frac{b}{c} = \frac{a-b}{c}$$

Simple Examples

$$\frac{4}{5} - \frac{2}{5} = \frac{4-2}{5} = \frac{2}{5}$$

$$\frac{5}{9} - \frac{3}{9} = \frac{2}{9}$$

$$\frac{11}{12} - \frac{5}{12} = \frac{6}{12} = \frac{1}{2}$$

Sample Problem 3-11

Find $\frac{5}{6} - \frac{2}{3}$.

Solution:

Step 1. Find the LCM of 6 and 3:
$$6 = (2)(3)$$
$$3 = \underline{(3)}$$
$$\text{LCD} = \text{LCM} = (2)(3) = 6$$

Step 2. Divide: $6 \div 6 = 1$ and $6 \div 3 = 2$

Step 3. Subtract: $\dfrac{5}{6} - \dfrac{2}{3} = \dfrac{\overbrace{5(1)}^{5}}{\underbrace{6(1)}_{6}} - \dfrac{\overbrace{2(2)}^{4}}{\underbrace{3(2)}_{6}} = \dfrac{1}{6}$

Sample Problem 3-12

Find $\frac{7}{12} - \frac{4}{15}$.

Solution:

Step 1. Find the LCM of 12 and 15:
$$12 = (2)(2)(3)$$
$$15 = \underline{(3)(5)}$$
$$\text{LCD} = \text{LCM} = (2)(2)(3)(5) = 60$$

Step 2. Divide: $60 \div 12 = 5$ and $60 \div 15 = 4$

Step 3. Subtract: $\dfrac{7}{12} - \dfrac{4}{15} = \dfrac{\overbrace{7(5)}^{35}}{\underbrace{12(5)}_{60}} - \dfrac{\overbrace{4(4)}^{16}}{\underbrace{15(4)}_{60}} = \dfrac{19}{60}$

Sample Problem 3-13

Find $\frac{31}{72} - \frac{19}{54}$.

Solution:

Step 1. Find the LCM of 72 and 54:
$$72 = (2)(2)(2)(3)(3)$$
$$54 = (2)(3)(3)(3)$$
$$\text{LCD} = \text{LCM} = (2)(2)(2)(3)(3)(3) = 216$$

Step 2. Divide: $216 \div 72 = 3$ and $216 \div 54 = 4$

Step 3. Subtract: $\dfrac{31}{72} - \dfrac{19}{54} = \dfrac{\overbrace{31(3)}^{93}}{\underbrace{72(3)}_{216}} - \dfrac{\overbrace{19(4)}^{76}}{\underbrace{54(4)}_{216}} = \dfrac{17}{216}$

Complete the problems below. Completed problems appear on page 106.

Sample Set

(a) $\dfrac{5}{8} - \dfrac{2}{8} = \dfrac{\Box}{\Box}$

(b) $\dfrac{8}{9} - \dfrac{2}{9} = \dfrac{\Box}{\Box} = \dfrac{\Box}{\Box}$

(c) $\dfrac{11}{18} - \dfrac{7}{12} = \dfrac{\Box}{36} - \dfrac{\Box}{36} = \dfrac{\Box}{\Box}$

(d) $\dfrac{7}{24} - \dfrac{5}{36} = \dfrac{\Box}{\Box} - \dfrac{\Box}{\Box} = \dfrac{\Box}{\Box}$

Justification of the Subtraction Rule

The subtraction rule can be justified in a manner similar to that used for the addition rule. The demonstration depends on the use of an equality similar to the distributive law.

$$C \times (A + B) = (C \times A) + (C \times B) \quad \textbf{distributive law}$$
$$C \times (A - B) = (C \times A) - (C \times B) \quad \textbf{needed equality}$$

The verification of the second equality depends on a knowledge of signed numbers, and it will be shown to hold in Unit 8. For the present, we shall accept the equality.

$$\frac{a}{c} - \frac{b}{c} = \left(a \times \frac{1}{c}\right) - \left(b \times \frac{1}{c}\right) \quad \textbf{definition of fractions}$$

$$= \left(\frac{1}{c} \times a\right) - \left(\frac{1}{c} \times b\right) \quad \textbf{commutative law of multiplication}$$

$$= \frac{1}{c} \times (a - b) \quad \textbf{equality}$$

$$= \frac{a - b}{c} \quad \textbf{definition of fraction}$$

Rules of Operation

multiplication rule	$\dfrac{a}{b} \times \dfrac{c}{d} = \dfrac{a \times c}{b \times d}$	
division rule	$\dfrac{a}{b} \div \dfrac{c}{d} = \dfrac{a}{b} \times \dfrac{d}{c}$	
addition rule	$\dfrac{a}{c} + \dfrac{b}{c} = \dfrac{a + b}{c}$	
subtraction rule	$\dfrac{a}{c} - \dfrac{b}{c} = \dfrac{a - b}{c}$	

Arithmetic of Common Fractions

Completed Problems

(a) $\dfrac{5}{8} - \dfrac{2}{8} = \boxed{\dfrac{3}{8}}$

(b) $\dfrac{8}{9} - \dfrac{2}{9} = \boxed{\dfrac{6}{9}} = \boxed{\dfrac{2}{3}}$

(c) $\dfrac{11}{18} - \dfrac{7}{12} = \dfrac{\boxed{22}}{36} - \dfrac{\boxed{21}}{36} = \dfrac{\boxed{1}}{\boxed{36}}$

(d) $\dfrac{7}{24} - \dfrac{5}{36} = \dfrac{\boxed{21}}{\boxed{72}} - \dfrac{\boxed{10}}{\boxed{72}} = \dfrac{\boxed{11}}{\boxed{72}}$

Name: _____

Class: _____

Exercises for Sec. 3-4 Find the differences. Express each answer in lowest terms.

1. $\dfrac{5}{7} - \dfrac{2}{7} =$
2. $\dfrac{3}{4} - \dfrac{1}{4} =$
3. $\dfrac{3}{4} - \dfrac{1}{2} =$

4. $\dfrac{5}{8} - \dfrac{3}{8} =$
5. $\dfrac{7}{6} - \dfrac{4}{6} =$
6. $\dfrac{2}{3} - \dfrac{1}{6} =$

7. $\dfrac{1}{2} - \dfrac{1}{3} =$
8. $\dfrac{1}{3} - \dfrac{1}{4} =$
9. $\dfrac{5}{9} - \dfrac{1}{6} =$

10. $\dfrac{7}{8} - \dfrac{3}{4} =$
11. $\dfrac{7}{9} - \dfrac{5}{12} =$
12. $\dfrac{11}{12} - \dfrac{7}{8} =$

13. $\dfrac{13}{18} - \dfrac{7}{12} =$ 14. $\dfrac{15}{20} - \dfrac{5}{8} =$ 15. $\dfrac{11}{15} - \dfrac{7}{12} =$

16. $\dfrac{3}{10} - \dfrac{2}{15} =$ 17. $\dfrac{5}{8} - \dfrac{7}{12} =$ 18. $\dfrac{13}{20} - \dfrac{7}{24} =$

19. $\dfrac{7}{28} - \dfrac{5}{21} =$ 20. $\dfrac{11}{24} - \dfrac{7}{18} =$ 21. $\dfrac{12}{16} - \dfrac{10}{24} =$

22. $\dfrac{13}{30} - \dfrac{7}{24} =$ 23. $\dfrac{19}{36} - \dfrac{11}{24} =$ 24. $\dfrac{5}{16} - \dfrac{2}{27} =$

25. $\dfrac{13}{18} - \dfrac{17}{27} =$

26. $\dfrac{11}{24} - \dfrac{5}{36} =$

27. $\dfrac{7}{50} - \dfrac{8}{75} =$

28. $\dfrac{17}{48} - \dfrac{5}{36} =$

29. $\dfrac{19}{72} - \dfrac{13}{54} =$

30. $\dfrac{7}{81} - \dfrac{5}{72} =$

Name: _____

Class: _____

Unit 3 Test Express each answer in lowest terms.

Find the products.

1. $\dfrac{2}{3} \times \dfrac{5}{7} =$
2. $\dfrac{3}{6} \times \dfrac{5}{7} =$
3. $\dfrac{5}{9} \times \dfrac{3}{10} =$

4. $\dfrac{12}{18} \times \dfrac{27}{16} =$
5. $\dfrac{36}{54} \times \dfrac{72}{64} =$

Find the quotients.

6. $\dfrac{3}{5} \div \dfrac{2}{7} =$
7. $\dfrac{5}{7} \div \dfrac{5}{8} =$
8. $\dfrac{\frac{3}{8}}{\frac{9}{4}} =$

9. $\dfrac{15}{18} \div \dfrac{27}{20} =$
10. $\dfrac{\frac{36}{56}}{\frac{48}{28}} =$

Find the sums.

11. $\dfrac{3}{7} + \dfrac{2}{7} =$
12. $\dfrac{5}{9} + \dfrac{3}{4} =$
13. $\dfrac{7}{12} + \dfrac{5}{18} =$

14. $\dfrac{17}{36} + \dfrac{13}{54} =$

15. $\dfrac{1}{12} + \dfrac{7}{18} + \dfrac{5}{24} =$

Find the differences.

16. $\dfrac{5}{13} - \dfrac{2}{13} =$

17. $\dfrac{5}{7} - \dfrac{3}{5} =$

18. $\dfrac{11}{12} - \dfrac{7}{15} =$

19. $\dfrac{13}{48} - \dfrac{7}{32} =$

20. $\dfrac{11}{54} - \dfrac{13}{72} =$

Section 4-1 Expressing Mixed Numbers as Improper Fractions

A **mixed number** is the sum of a whole number and a common fraction.

$$4\tfrac{1}{2} = 4 + \tfrac{1}{2}$$
$$5\tfrac{2}{3} = 5 + \tfrac{2}{3}$$
$$95\tfrac{7}{8} = 95 + \tfrac{7}{8}$$

Mixed numbers are equivalent to improper fractions. Using the mixed number $4\tfrac{2}{3}$, this is demonstrated below.

$$4\tfrac{2}{3} = 4 + \tfrac{2}{3}$$

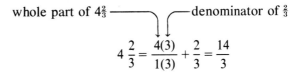

$$= \tfrac{4}{1} + \tfrac{2}{3}$$

$$= \tfrac{4(3)}{1(3)} + \tfrac{2}{3}$$

$$= \tfrac{12}{3} + \tfrac{2}{3} = \tfrac{14}{3}$$

The key step was to write $4 = \tfrac{4}{1}$ as an equivalent fraction with the same denominator as the fractional part $\tfrac{2}{3}$. A shortcut can be seen when the demonstration is written with only this key step shown:

whole part of $4\tfrac{2}{3}$ — denominator of $\tfrac{2}{3}$

$$4\tfrac{2}{3} = \tfrac{4(3)}{1(3)} + \tfrac{2}{3} = \tfrac{14}{3}$$

Numerator of improper fraction = whole part × denominator + numerator
of $4\tfrac{2}{3}$ of $\tfrac{2}{3}$ of $\tfrac{2}{3}$

(Example continues on the next page.)

Complete the problems below. Completed problems appear on page 116.

Sample Set

(a) $5\tfrac{3}{4} = \tfrac{5()}{1(\,4\,)} + \tfrac{3}{4} = \tfrac{\Box + 3}{4} = \tfrac{\Box}{4}$

(b) $7\tfrac{2}{5} = \tfrac{\Box + 2}{5} = \tfrac{\Box}{5}$

(c) $6\tfrac{5}{8} = \tfrac{\Box}{8}$

Thus,
$$4 + \frac{2}{3} = \frac{4(3) + 2}{3}$$

Notice that the denominator of the improper fraction obtained is the same as the denominator of the fractional part of the mixed number.

It is important to realize that mixed numbers are fractions. This follows from the fact that they are equivalent to improper fractions.

In general,
$$a + \frac{b}{c} = \frac{ac + b}{c}$$

where a, b, and c are numbers, and $c \neq 0$.

Sample Problem 4-1 Express $23\frac{2}{3}$ as an improper fraction.

Solution:
$$23 \times 3 + 2 = 69 + 2 = 71$$
$$23\tfrac{2}{3} = \tfrac{71}{3} \qquad \text{answer}$$

Completed Problems

(a) $\quad 5\dfrac{3}{4} = \dfrac{5(\boxed{4})}{1(\,4\,)} + \dfrac{3}{4} = \dfrac{\boxed{20} + 3}{4} = \dfrac{\boxed{23}}{4}$

(b) $\quad 7\dfrac{2}{5} = \dfrac{\boxed{35} + 2}{5} = \dfrac{\boxed{37}}{5}$

(c) $\quad 6\dfrac{5}{8} = \dfrac{\boxed{53}}{8}$

Name: _____

Class: _____

Exercises for Sec. 4-1 Write each mixed number as an improper fraction.

1. $2\frac{1}{2} =$
2. $3\frac{1}{2} =$

3. $3\frac{2}{3} =$
4. $5\frac{3}{5} =$

5. $6\frac{6}{7} =$
6. $7\frac{3}{8} =$

7. $4\frac{7}{8} =$
8. $5\frac{2}{9} =$

9. $3\frac{7}{10} =$
10. $8\frac{3}{4} =$

11. $14\frac{1}{3} =$
12. $23\frac{2}{3} =$

13. $15\frac{3}{4} =$
14. $17\frac{5}{8} =$

15. $32\frac{3}{5} =$
16. $43\frac{2}{5} =$

17. $53\frac{1}{2} =$
18. $83\frac{2}{9} =$

19. $62\frac{5}{6} =$
20. $78\frac{3}{10} =$

21. $3\frac{5}{12} =$
22. $5\frac{2}{15} =$

23. $6\frac{3}{10} =$
24. $2\frac{7}{18} =$

25. $7\frac{5}{24} =$ 26. $8\frac{7}{36} =$

27. $1\frac{3}{20} =$ 28. $3\frac{3}{32} =$

29. $4\frac{7}{54} =$ 30. $9\frac{11}{72} =$

Section 4-2 Expressing Improper Fractions as Mixed Numbers

The long-division process can be used to find a mixed number equivalent to an improper fraction. For example, to express $\frac{548}{37}$ as a mixed number, divide 548 by 37 using long division.

$$
\begin{array}{r}
14\frac{30}{37} \\
37\overline{)548} \\
\underline{37} \\
178 \\
\underline{148} \\
30
\end{array}
$$

The mixed number $14\frac{30}{37}$ can be shown equivalent to $\frac{548}{37}$ using the same procedure as was used at the beginning of the preceding section.

$$14\frac{30}{37} = 14 + \frac{30}{37}$$

$$= \frac{14}{1} + \frac{30}{37}$$

$$= \frac{14(37)}{1(37)} + \frac{30}{37}$$

$$= \frac{518}{37} + \frac{30}{37} = \frac{548}{37}$$

Another example follows on page 120.

Complete each problem below. Completed problems appear on page 120.

Sample Set

(a) $\quad \frac{8}{3} = 2\frac{2}{\Box}$

(b) $\quad \frac{15}{4} = 3\frac{\Box}{4}$

(c) $\quad \frac{7}{2} = \Box\frac{1}{2}$

(d) $\quad \frac{29}{5} = \Box\frac{\Box}{5}$

(e) $\quad \frac{47}{13} = \Box\frac{\Box}{\Box}$

Sample Problem 4-2   Express $\frac{392}{53}$ as a mixed number.

Solution:

$$53 \overline{\smash{\big)}\, 392}^{\,7\frac{21}{53}}$$
$$\underline{371}$$
$$21$$

$\frac{392}{53} = 7\frac{21}{53}$

Completed Problems

(a) $\dfrac{8}{3} = 2\dfrac{2}{\boxed{3}}$

(b) $\dfrac{15}{4} = 3\dfrac{\boxed{3}}{4}$

(c) $\dfrac{7}{2} = \boxed{3}\dfrac{1}{2}$

(d) $\dfrac{29}{5} = \boxed{5}\dfrac{\boxed{4}}{5}$

(e) $\dfrac{47}{13} = \boxed{3}\dfrac{\boxed{8}}{\boxed{13}}$

Name: _____

Class: _____

Exercises for Sec. 4-2 Express each improper fraction as a mixed number. Before doing this, make sure that each improper fraction is in lowest terms.

1. $\frac{3}{2} =$ 2. $\frac{5}{3} =$ 3. $\frac{7}{2} =$

4. $\frac{7}{5} =$ 5. $\frac{5}{4} =$ 6. $\frac{6}{4} =$

7. $\frac{8}{3} =$ 8. $\frac{9}{5} =$ 9. $\frac{6}{5} =$

10. $\frac{8}{6} =$ 11. $\frac{12}{7} =$ 12. $\frac{15}{10} =$

13. $\frac{17}{3} =$ 14. $\frac{13}{4} =$ 15. $\frac{18}{10} =$

16. $\frac{19}{4} =$ 17. $\frac{17}{6} =$ 18. $\frac{11}{10} =$

19. $\frac{16}{3} =$ 20. $\frac{22}{3} =$ 21. $\frac{24}{15} =$

22. $\frac{34}{7} =$ 23. $\frac{25}{13} =$ 24. $\frac{29}{15} =$

25. $\frac{32}{12} =$ 26. $\frac{53}{4} =$ 27. $\frac{69}{2} =$

28. $\frac{73}{7} =$ 29. $\frac{64}{7} =$ 30. $\frac{75}{18} =$

31. $\frac{84}{15} =$ 32. $\frac{69}{8} =$ 33. $\frac{125}{3} =$

34. $\frac{212}{6} =$ 35. $\frac{513}{7} =$ 36. $\frac{436}{7} =$

37. $\frac{59}{13} =$ 38. $\frac{72}{11} =$ 39. $\frac{83}{29} =$

40. $\frac{94}{48} =$ 41. $\frac{77}{23} =$ 42. $\frac{87}{31} =$

43. $\frac{56}{17} =$ 44. $\frac{69}{24} =$ 45. $\frac{173}{51} =$

46. $\frac{261}{23} =$ 47. $\frac{547}{95} =$ 48. $\frac{378}{37} =$

49. $\frac{862}{173} =$ 50. $\frac{786}{281} =$

Section 4-3 Multiplication and Division of Mixed Numbers

When multiplying mixed numbers, multiply their equivalent improper fractions.

Sample Problem 4-3

Multiply $4\frac{1}{5} \times 2\frac{7}{9}$.

Solution:

Write each mixed number as an improper fraction; then multiply:

$$4\frac{1}{5} \times 2\frac{7}{9} = \frac{\cancel{21}^{7}}{\cancel{5}_{1}} \times \frac{\cancel{25}^{5}}{\cancel{9}_{3}} = \frac{35}{3} = 11\frac{2}{3}$$

Sample Problem 4-4

Multiply $5\frac{1}{3} \times \frac{2}{15} \times 5\frac{1}{4}$.

Solution:

$$5\frac{1}{3} \times \frac{2}{15} \times 5\frac{1}{4} = \frac{\cancel{16}^{4}}{\cancel{3}_{1}} \times \frac{2}{15} \times \frac{\cancel{21}^{7}}{\cancel{4}_{1}} = \frac{56}{15} = 3\frac{11}{15}$$

Sample Problem 4-5

Multiply $8\frac{1}{3} \times \frac{2}{5} \times 3$.

Solution:

$$8\frac{1}{3} \times \frac{2}{5} \times 3 = \frac{\cancel{25}^{5}}{\cancel{3}_{1}} \times \frac{2}{\cancel{5}_{1}} \times \frac{\cancel{3}^{1}}{1} = \frac{10}{1} = 10$$

Sample problems continue on page 124.

Complete each problem below. Completed problems appear on page 124.

Sample Set

(a) $3\frac{1}{2} \times 1\frac{5}{7} = \frac{\boxed{}}{\cancel{2}_{1}} \times \frac{\boxed{}}{\cancel{7}_{1}} = \boxed{}$

(b) $3\frac{3}{5} \times 1\frac{1}{6} \times 1\frac{4}{11} = \frac{\boxed{}}{\cancel{5}_{1}} \times \frac{\boxed{}}{\cancel{6}_{1}} \times \frac{\boxed{}}{11} = \frac{\boxed{}}{\boxed{}} = \boxed{}$

(c) $17\frac{1}{2} \div 10\frac{1}{5} = \frac{\boxed{}}{2} \div \frac{\boxed{}}{5} = \frac{\boxed{}}{2} \times \frac{5}{\boxed{}} = \frac{\boxed{}}{\boxed{}}$
$= \boxed{}$

In division, as in multiplication, it is necessary to first express each mixed number as an improper fraction.

Sample Problem 4-6 Divide $3\frac{3}{5} \div 2\frac{7}{10}$.

Solution:
$$3\frac{3}{5} \div 2\frac{7}{10} = \frac{18}{5} \div \frac{27}{10} = \frac{\cancel{18}^2}{\cancel{5}_1} \times \frac{\cancel{10}^2}{\cancel{27}_3} = \frac{4}{3}$$

Completed Problems

(a) $3\frac{1}{2} \times 1\frac{5}{7} = \dfrac{\cancel{7}^1}{\cancel{2}_1} \times \dfrac{\cancel{12}^6}{\cancel{7}_1} = \boxed{6}$

(b) $3\frac{3}{5} \times 1\frac{1}{6} \times 1\frac{4}{11} = \dfrac{\cancel{18}^3}{\cancel{5}_1} \times \dfrac{\cancel{7}}{\cancel{6}_1} \times \dfrac{\cancel{15}^3}{11} = \boxed{\dfrac{63}{11}} = \boxed{5\dfrac{8}{11}}$

(c) $17\frac{1}{2} \div 10\frac{1}{5} = \dfrac{35}{2} \div \dfrac{51}{5} = \dfrac{35}{2} \times \dfrac{5}{\boxed{51}} = \boxed{\dfrac{175}{102}}$

$\qquad = \boxed{1\dfrac{73}{102}}$

Name: _____

Class: _____

Exercises for Sec. 4-3 Find the products. Express your answers as mixed numbers with fractional parts in lowest terms.

1. $4\frac{1}{2} \times 2\frac{2}{3} =$

2. $1\frac{3}{4} \times 2\frac{2}{5} =$

3. $2\frac{1}{2} \times 1\frac{1}{3} =$

4. $1\frac{1}{5} \times 3\frac{3}{4} =$

5. $2\frac{2}{3} \times 2\frac{1}{2} =$

6. $1\frac{4}{5} \times 1\frac{2}{9} =$

7. $6\frac{1}{4} \times 1\frac{1}{5} =$

8. $2\frac{2}{7} \times 1\frac{1}{2} =$

9. $2\frac{4}{7} \times 3\frac{2}{3} =$

10. $3\frac{2}{5} \times 7\frac{1}{2} =$

11. $2\frac{7}{9} \times 1\frac{7}{20} =$

12. $9\frac{3}{5} \times 2\frac{7}{9} =$

13. $5\frac{1}{7} \times 5\frac{1}{4} =$

14. $7\frac{5}{7} \times 7\frac{7}{8} =$

15. $4\frac{4}{11} \times 2\frac{4}{9} =$

16. $7\frac{1}{9} \times 2\frac{7}{10} =$

17. $12\frac{3}{5} \times 1\frac{7}{18} =$

18. $2\frac{22}{25} \times 2\frac{2}{9} =$

19. $3\frac{11}{15} \times 1\frac{4}{21} =$

20. $5\frac{1}{16} \times 1\frac{13}{27} =$

21. $1\frac{1}{3} \times 2\frac{1}{2} \times 1\frac{4}{5} =$

22. $7\frac{1}{2} \times \frac{1}{5} \times 1\frac{1}{3} =$

23. $5\frac{1}{3} \times 1\frac{1}{4} \times \frac{7}{20} =$

24. $3\frac{3}{8} \times 1\frac{1}{12} \times 1\frac{3}{13} =$

25. $1\frac{2}{9} \times 1\frac{1}{2} \times 1\frac{1}{5} =$

26. $7\frac{1}{5} \times 1\frac{1}{8} \times 2\frac{1}{7} =$

27. $2\frac{2}{5} \times 1\frac{2}{7} \times 3\frac{1}{3} =$

28. $1\frac{3}{4} \times 1\frac{11}{14} \times \frac{9}{15} =$

29. $1\frac{11}{16} \times 3\frac{11}{15} \times 1\frac{4}{21} =$

30. $3\frac{3}{8} \times 2\frac{7}{9} \times 1\frac{1}{15} =$

Find the quotients. Express your answers as proper fractions or mixed numbers in lowest terms.

31. $4\frac{1}{2} \div 1\frac{1}{5} =$ 32. $1\frac{1}{5} \div 1\frac{1}{2} =$

33. $3\frac{1}{2} \div 1\frac{3}{4} =$ 34. $2\frac{2}{3} \div 1\frac{1}{5} =$

35. $2\frac{1}{3} \div 1\frac{2}{3} =$ 36. $1\frac{1}{6} \div \frac{2}{9} =$

37. $2\frac{2}{5} \div 2 =$ 38. $1\frac{5}{6} \div 1\frac{2}{9} =$

39. $7\frac{1}{2} \div 6\frac{2}{3} =$ 40. $3\frac{1}{4} \div 8\frac{2}{3} =$

41. $2\frac{2}{9} \div 2\frac{1}{12} =$ 42. $4\frac{4}{5} \div 2\frac{2}{15} =$

43. $5\frac{1}{7} \div 6\frac{6}{7} =$ 44. $6\frac{6}{7} \div 10\frac{2}{3} =$

45. $3\frac{1}{9} \div 2\frac{13}{18} =$ 46. $5\frac{1}{7} \div 13\frac{1}{2} =$

47. $9\frac{3}{5} \div 1\frac{1}{15} =$

48. $9\frac{1}{7} \div 48 =$

49. $3\frac{1}{3} \div 2\frac{2}{3} =$

50. $6\frac{2}{9} \div 8\frac{1}{6} =$

51. $2\frac{3}{16} \div 2\frac{1}{24} =$

52. $2\frac{13}{16} \div 3\frac{1}{8} =$

53. $1\frac{11}{24} \div 1\frac{13}{15} =$

54. $1\frac{19}{35} \div 1\frac{13}{14} =$

55. $2\frac{2}{35} \div 1\frac{11}{25} =$

Section 4-4 Addition of Mixed Numbers

The sum of mixed numbers is most conveniently computed by adding the whole parts and fractional parts separately.

$$\begin{array}{r} 5\frac{2}{7} \\ +3\frac{4}{7} \\ \hline 8\frac{6}{7} \end{array} \qquad \begin{array}{l} \text{Add columns:} \\ \frac{2}{7} + \frac{4}{7} = \frac{6}{7} \\ 5 + 3 = 8 \end{array}$$

Can you guess what laws are used when you add whole and fractional parts separately?

$$\begin{aligned} 5\tfrac{2}{7} + 3\tfrac{4}{7} &= (5 + \tfrac{2}{7}) + (3 + \tfrac{4}{7}) \\ &= (5 + 3) + (\tfrac{2}{7} + \tfrac{4}{7}) \\ &= 8 + \tfrac{6}{7} = 8\tfrac{6}{7} \end{aligned}$$

the commutative and associative laws are used here to rearrange and regroup numbers.

Sample Problem 4-7 Add $17\frac{8}{9} + 24\frac{5}{9}$.

Solution:

$$\begin{array}{r} 17\frac{8}{9} \\ +24\frac{5}{9} \\ \hline 41\frac{13}{9} \end{array} = 41 + 1\tfrac{4}{9} = 42\tfrac{4}{9}$$

Here, $\frac{13}{9} = 1\frac{4}{9}$.

Sample Problems continue on page 130.

Complete the problems below. Completed problems appear on page 131.

Sample Set

(a) $3\dfrac{3}{8} = 3\dfrac{\square}{24}$

$+5\dfrac{1}{6} = +5\dfrac{\square}{24}$

$\dfrac{\square}{24}$

(b) $5\dfrac{11}{18} = 5\dfrac{\square}{72}$

$+7\dfrac{13}{24} = +7\dfrac{\square}{72}$

$\dfrac{\square}{72} = \square\dfrac{\square}{72}$

(c) $6\dfrac{7}{9} = 6\dfrac{\square}{36}$

$7\dfrac{11}{12} = 7\dfrac{\square}{36}$

$+4\dfrac{13}{18} = +4\dfrac{\square}{36}$

$17\dfrac{\square}{36} = \square\dfrac{\square}{36} = \square\dfrac{\square}{\square}$

Sample Problem 4-8 Add $5\frac{7}{12} + 3\frac{5}{18}$.

Solution: The LCM of 12 and 18 is 36; write each fractional part with LCD = 36: $\frac{7}{12} = \frac{21}{36}$ $\frac{5}{18} = \frac{10}{36}$

You would write the problem as

$$\begin{array}{r} 5\frac{7}{12} = & 5\frac{21}{36} \\ +3\frac{5}{18} = & +3\frac{10}{36} \\ \hline & 8\frac{31}{36} \end{array}$$

Sample Problem 4-9 Add $6\frac{5}{8} + 3\frac{11}{18} + 7\frac{13}{24}$.

Solution: The LCM of 8, 18, and 24 is 72.

$$\frac{5}{8} = \frac{45}{72} \qquad \frac{11}{18} = \frac{44}{72} \qquad \frac{13}{24} = \frac{39}{72}$$

You should write the problem as

$$\begin{array}{r} 6\frac{5}{8} = & 6\frac{45}{72} \\ 3\frac{11}{18} = & 3\frac{44}{72} \\ +7\frac{13}{24} = & +7\frac{39}{72} \\ \hline & 16\frac{128}{72} = 17\frac{56}{72} = 17\frac{7}{9} \end{array}$$

Scratch work:

$$\begin{array}{r} 45 \\ 44 \\ 39 \\ \hline 128 \end{array} \qquad \begin{array}{r} 1 \\ 72\overline{)128} \\ \underline{72} \\ 56 \end{array} \qquad \begin{array}{r} 7 \\ \frac{56}{72} \\ 9 \end{array}$$

Completed Problems

(a) $\quad 3\frac{3}{8} = \quad 3\frac{\boxed{9}}{24}$

$\quad +5\frac{1}{6} = +5\frac{\boxed{4}}{24}$

$\quad\quad\quad\quad\quad \boxed{8}\frac{\boxed{13}}{24}$

(b) $\quad 5\frac{11}{18} = \quad 5\frac{\boxed{44}}{72}$

$\quad +7\frac{13}{24} = +7\frac{\boxed{39}}{72}$

$\quad\quad\quad\quad \boxed{12}\frac{\boxed{83}}{72} = \boxed{13}\frac{\boxed{11}}{72}$

(c) $\quad 6\frac{7}{9} = \quad 6\frac{\boxed{28}}{36}$

$\quad\quad 7\frac{11}{12} = \quad 7\frac{\boxed{33}}{36}$

$\quad +4\frac{13}{18} = +4\frac{\boxed{26}}{36}$

$\quad\quad\quad\quad 17\frac{\boxed{87}}{36} = \boxed{19}\frac{\boxed{15}}{36} = \boxed{19}\frac{\boxed{5}}{\boxed{12}}$

Name: _____

Class: _____

Exercises for Sec. 4-4 Find the sums. Express each answer in lowest terms.

1. $2\frac{1}{3}$
   $+4\frac{2}{3}$

2. $3\frac{1}{4}$
   $+5\frac{2}{4}$

3. $7\frac{3}{8}$
   $+6\frac{3}{8}$

4. $5\frac{5}{6}$
   $+4\frac{3}{6}$

5. $4\frac{1}{2}$
   $+2\frac{1}{3}$

6. $8\frac{1}{3}$
   $+4\frac{3}{4}$

7. $7\frac{7}{9}$
   $+8\frac{1}{3}$

8. $9\frac{5}{8}$
   $+2\frac{1}{4}$

9. $7\frac{7}{12}$
   $+8\frac{11}{18}$

10. $13\frac{11}{15}$
    $+14\frac{5}{12}$

11. $15\frac{11}{18}$
    $+16\frac{5}{24}$

12. $16\frac{7}{15}$
    $+\ 9\frac{3}{20}$

13. $32\frac{9}{20}$
    $+43\frac{8}{25}$

14. $27\frac{9}{28}$
    $+64\frac{10}{21}$

15. $41\frac{7}{16}$
    $+53\frac{13}{24}$

16. $81\frac{11}{15}$
    $+47\frac{13}{25}$

17. $29\frac{7}{12}$
    $+53\frac{9}{16}$

18. $47\frac{13}{24}$
    $+29\frac{15}{36}$

19. $78\frac{17}{21}$
    $+37\frac{20}{49}$

20. $47\frac{19}{36}$
    $+29\frac{25}{48}$

21. $67\frac{19}{36}$
    $+19\frac{31}{45}$

22. $45\frac{17}{20}$
    $+72\frac{23}{30}$

23. $31\frac{17}{32}$
    $+89\frac{23}{48}$

24. $47\frac{11}{54}$
    $+84\frac{13}{36}$

25. $63\frac{11}{48}$
    $+79\frac{7}{54}$

26. $7\frac{1}{4}$
    $6\frac{3}{4}$
    $+5\frac{1}{2}$

27. $5\frac{4}{9}$
    $1\frac{5}{6}$
    $+4\frac{1}{4}$

28. $4\frac{5}{6}$
    $3\frac{3}{4}$
    $+5\frac{2}{3}$

29. $4\frac{5}{6}$
    $5\frac{3}{8}$
    $+9\frac{1}{12}$

30. $7\frac{1}{2}$
    $6\frac{2}{3}$
    $+5\frac{2}{5}$

31. $5\frac{7}{10}$
    $8\frac{5}{6}$
    $+2\frac{2}{15}$

32. $5\frac{5}{12}$
    $3\frac{3}{8}$
    $+\frac{7}{18}$

33. $9\frac{3}{20}$
    $7\frac{8}{15}$
    $+8\frac{7}{12}$

34. $7\frac{11}{18}$
    $8\frac{7}{24}$
    $+1\frac{5}{12}$

35. $2\frac{11}{20}$
    $5\frac{9}{10}$
    $+4\frac{12}{25}$

36. $7\frac{7}{24}$
    $12\frac{5}{36}$
    $+13\frac{7}{18}$

37. $18\frac{2}{21}$
$12\frac{6}{49}$
$+\ 17\frac{3}{28}$

38. $84\frac{7}{36}$
$37\frac{5}{48}$
$+\ 29\frac{3}{16}$

39. $32\frac{11}{42}$
$15\frac{9}{28}$
$+\ 29\frac{8}{21}$

40. $12\frac{5}{72}$
$37\frac{7}{81}$
$+\ 18\frac{3}{64}$

Section 4-5 Subtraction of Mixed Numbers

The difference between two mixed numbers can be found by subtracting the whole and fractional parts separately.

$$\begin{array}{r} 8\frac{7}{8} \\ -3\frac{3}{8} \\ \hline 5\frac{4}{8} = 5\frac{1}{2} \end{array}$$

Subtract columns:
$\frac{7}{8} - \frac{3}{8} = \frac{4}{8} = \frac{1}{2}$
$8 - 3 = 5$

Sample Problem 4-10

Solution:

Subtract $7\frac{7}{12} - 4\frac{4}{15}$.

The LCM of 12 and 15 is 60.

$$\frac{7}{12} = \frac{35}{60} \qquad \frac{4}{15} = \frac{16}{60}$$

Then write the problem:

$$\begin{array}{r} 7\frac{7}{12} = 7\frac{35}{60} \\ -4\frac{4}{15} = -4\frac{16}{60} \\ \hline 3\frac{19}{60} \end{array}$$

The example below demonstrates a case in which 1 has to be borrowed.

$$\begin{array}{r} 7\frac{2}{9} \\ -5\frac{5}{9} \\ \hline \end{array}$$

(Example continues on page 138.)

Complete the problems below. Completed problems appear on page 139.

Sample Set

(a) $\quad 5\frac{5}{8}$
$\quad -2\frac{3}{8}$
$\quad \boxed{}\frac{\boxed{}}{8} = \boxed{}\frac{\boxed{}}{\boxed{}}$

(b) $\quad 5\frac{11}{18} = 5\frac{\boxed{}}{36}$
$\quad -3\frac{7}{12} = -3\frac{\boxed{}}{36}$
$\quad \boxed{}\frac{\boxed{}}{36}$

(c) $\quad 7\frac{11}{24} = 7\frac{\boxed{}}{72} = \boxed{}\frac{\boxed{}}{72}$
$\quad -6\frac{19}{36} = -6\frac{\boxed{}}{72} = -6\frac{\boxed{}}{72}$
$\quad \phantom{-6\frac{19}{36} = -6\frac{\boxed{}}{72} =} \frac{\boxed{}}{72}$

138 Mixed-Number Arithmetic

Here the fractional part $\frac{2}{9}$ is less than $\frac{5}{9}$. It is, therefore, necessary to borrow 1 (really $\frac{9}{9}$) from 7.

$$
\begin{aligned}
7\tfrac{2}{9} &= 7 + \tfrac{2}{9} \\
&= (6 + 1) + \tfrac{2}{9} \\
&= (6 + \tfrac{9}{9}) + \tfrac{2}{9} \\
&= 6 + (\tfrac{9}{9} + \tfrac{2}{9}) \qquad \textbf{what law?} \\
&= 6 + \tfrac{11}{9} = 6\tfrac{11}{9}
\end{aligned}
$$

There is a shortcut that allows us to write

$$7\frac{2}{9} \quad \text{as} \quad 6\frac{11}{9} \quad \leftarrow 2 + 9$$

in one step. Notice that the numerator 11 is the sum of the numerator and denominator of $\frac{2}{9}$.

New numerator = old numerator + denominator

The problem is completed below as you would write it.

$$
\begin{array}{r}
7\tfrac{2}{9} = 6\tfrac{11}{9} \\
-5\tfrac{5}{9} = -5\tfrac{5}{9} \\ \hline
1\tfrac{6}{9} = 1\tfrac{2}{3}
\end{array}
$$

Sample Problem 4-11

Solution:

Subtract $8\frac{8}{15} - 7\frac{11}{12}$.

The LCM of 15 and 12 is 60.

$$\tfrac{8}{15} = \tfrac{32}{60} \qquad \tfrac{11}{12} = \tfrac{55}{60}$$

Then write the problem:

$$
\begin{array}{r}
8\tfrac{8}{15} = 8\tfrac{32}{60} = 7\tfrac{92}{60} \\
-7\tfrac{11}{12} = -7\tfrac{55}{60} = -7\tfrac{55}{60} \\ \hline
\tfrac{37}{60}
\end{array}
$$

There is no point in writing 0 here.

Sec. 4-5 Subtraction of Mixed Numbers

Completed Problems

(a) $5\frac{5}{8}$

$-2\frac{3}{8}$

$\boxed{3}\frac{\boxed{2}}{8} = \boxed{3}\frac{\boxed{1}}{4}$

(b) $5\frac{11}{18} = 5\frac{\boxed{22}}{36}$

$-3\frac{7}{12} = -3\frac{\boxed{21}}{36}$

$\phantom{-3\frac{7}{12} = -}\boxed{2}\frac{\boxed{1}}{36}$

(c) $7\frac{11}{24} = 7\frac{\boxed{33}}{72} = \boxed{6}\frac{\boxed{105}}{72}$

$-6\frac{19}{36} = -6\frac{\boxed{38}}{72} = -6\frac{\boxed{38}}{72}$

$\phantom{-6\frac{19}{36} = -6\frac{38}{72} = -6}\frac{\boxed{67}}{72}$

Name: _____

Class: _____

Exercises for Sec. 4-5   Find the differences.   Express each answer in lowest terms.

1. $5\frac{3}{4}$
     $-4\frac{1}{4}$

2. $7\frac{7}{10}$
     $-3\frac{4}{10}$

3. $7\frac{11}{13}$
     $-4\frac{7}{13}$

4. $8\frac{3}{8}$
     $-2\frac{7}{8}$

5. $6\frac{2}{7}$
     $-5\frac{5}{7}$

6. $9\frac{2}{3}$
     $-6\frac{1}{2}$

7. $8\frac{1}{4}$
     $-5\frac{1}{2}$

8. $6\frac{5}{6}$
     $-4\frac{3}{4}$

9. $15\frac{7}{9}$
     $-\ 9\frac{5}{6}$

10. $23\frac{7}{8}$
     $-17\frac{3}{4}$

11. $15\frac{7}{10}$
     $-\ 7\frac{7}{12}$

12. $28\frac{7}{12}$
     $-13\frac{9}{16}$

13. $43\frac{7}{15}$
     $-35\frac{5}{9}$

14. $67\frac{17}{18}$
     $-25\frac{11}{12}$

15. $73\frac{7}{15}$
     $-24\frac{3}{20}$

16. $81\frac{7}{18}$
     $-54\frac{5}{24}$

17. $18\frac{7}{18}$
     $-17\frac{4}{21}$

18. $21\frac{13}{15}$
     $-16\frac{7}{25}$

19. $55\frac{11}{20}$
    $-43\frac{5}{12}$

20. $24\frac{9}{14}$
    $-17\frac{4}{21}$

21. $57\frac{17}{24}$
    $-28\frac{13}{20}$

22. $47\frac{7}{24}$
    $-32\frac{7}{36}$

23. $43\frac{19}{48}$
    $-28\frac{13}{36}$

24. $57\frac{11}{40}$
    $-21\frac{13}{32}$

25. $86\frac{19}{50}$
    $-24\frac{7}{25}$

26. $37\frac{11}{54}$
    $-19\frac{13}{72}$

27. $56\frac{8}{27}$
    $-29\frac{5}{36}$

28. $47\frac{7}{30}$
    $-27\frac{3}{20}$

29. $28\frac{7}{56}$
    $-13\frac{8}{49}$

30. $74\frac{5}{72}$
    $-39\frac{7}{81}$

Name: _____

Class: _____

Applications
Multiplication and Division

1. If a man drives at a speed of 60 miles per hour, how far will he travel in $3\frac{1}{5}$ hours?
   (*Hint:* How far would he go in 3 hours?)

2. A dealer sells sand for 15 dollars per cubic yard. How much should $2\frac{1}{3}$ cubic yards cost?
   (*Hint:* How much would 2 cubic yards cost?)

3. A truck can carry $2\frac{1}{2}$ cubic yards of sand per load. How many trips will be required to carry $12\frac{1}{2}$ cubic yards?
   (*Hint:* How many trips are required if you want 12 cubic yards and can haul 2 cubic yards per trip?)

4. If flour sells for 54 cents per pound, how much would the following cost:
   (*a*) $\frac{2}{3}$ pounds?
   (*b*) $3\frac{1}{2}$ pounds?
   (*Hint:* How much would 3 pounds cost?)

5. A manufacturer claims that a gallon of his paint will cover 500 square feet. How many square feet would $8\frac{1}{5}$ gallons cover?
   (*Hint:* How many square feet would 8 gallons cover?)

6. If a gallon of paint covers 500 square feet, what fraction of a gallon is required to cover $16\frac{2}{3}$ square feet?
(*Hint:* What fraction of a gallon is required to cover 250 square feet? 100 square feet? 20 square feet?)

7. Joe buys $2\frac{1}{2}$ pounds of hamburger and gives one-quarter of it to John. How many pounds of hamburger did he give John?
(*Hint:* What operation does the word *of* indicate? What is one-half of 2? One-quarter of 2?)

8. A patient has been receiving $\frac{1}{16}$ grain of morphine. The doctor decides that the patient is better and tells the nurse to give the patient half as much. How much of a grain should the nurse give?
(*Hint:* What is one-half of $\frac{1}{16}$? What operation does the word *of* indicate?)

9. A snail crawls $2\frac{1}{3}$ inches per minute. How many minutes will it take the snail to crawl $22\frac{1}{2}$ inches?
(*Hint:* How long will it take the snail to crawl 22 inches if it crawls 2 inches per minute?)

10. A woman drove 390 miles in $7\frac{1}{2}$ hours.
    (*a*) What was her average speed?
    (*Hint:* How fast would she drive if she traveled 400 miles in 8 hours?)
    (*b*) At that rate, how far did she drive in $4\frac{1}{4}$ hours?
    (*Hint:* How far in 4 hours?)

11. A person can assemble a bicycle in $\frac{3}{4}$ hour. How long will it take the person to assemble 5 bicycles?
    (*Hint:* How long if one bicycle every 2 hours? One per hour?)

12. An item is sold in $7\frac{1}{2}$-pound units. How many units must be purchased in order to have 45 pounds?
    (*Hint:* How many units if each weighs 9 pounds?)

13. A certain kind of cloth costs $\$2\frac{1}{2}$ per yard. How much would $1\frac{3}{5}$ yards cost?
    (*Hint:* How much for 2 yards if $2 per yard?)

14. A man can do a job in $3\frac{1}{2}$ hours. If he works for 2 hours and quits, what fraction of the job did he do?
    (*Hint:* What fraction if he worked 2 hours on a 4-hour job?)

15. If a pump can empty a tank at a rate of $7\frac{1}{4}$ gallons per minute, how many gallons will be emptied in $3\frac{1}{2}$ minutes?
    (*Hint:* If 7 gallons per minute, how many gallons in 3 minutes?)

16. A tank containing 500 gallons is to be emptied at a rate of $6\frac{2}{3}$ gallons per minute. How long will it take to empty the tank?
    (*Hint:* How long to empty 500 gallons if you remove 5 gallons per minute?)

17. When small sea animals eat algae, their bodies retain $\frac{3}{20}$ of the calories taken in. When a smelt eats the small sea animals, its body retains $\frac{1}{5}$ of those calories. If a man eats the smelt, he retains $\frac{1}{5}$ of the food calories in the smelt. What fraction of the original calories in the algae would a man retain?

Addition and Subtraction

18. A homemaker buys $2\frac{1}{2}$ pounds of pork chops, $1\frac{3}{4}$ pounds of steak, and 2 pounds of hamburger. How much meat has been purchased?

19. A 2-pound sample of hamburger was found to contain $1\frac{2}{3}$ pounds of pure beef, and the rest was cereal. How much cereal was in the hamburger?

20. A can contains $\frac{7}{8}$ gallon of gasoline. If $\frac{1}{2}$ gallon is poured out, how much gasoline remains in the can?

21. On the days that Mary studies, she spends $1\frac{1}{2}$ hours studying English, $\frac{3}{4}$ hour studying history, and $\frac{1}{5}$ hour studying mathematics. How much total time does she spend studying on one of those days?

22. A metal plate, $\frac{1}{4}$ inch thick, is plated on one side with silver. The plate is then found to be $\frac{67}{128}$ inch thick. How thick is the silver plating?

23. A metal plate $\frac{5}{32}$ inch thick is dipped in a coating solution that covers each side with a coating that is $\frac{1}{128}$ inch thick. How thick is the coated plate?

24. A floor is made by laying $\frac{3}{4}$-inch oak flooring over a layer of $1\frac{1}{2}$-inch-thick subflooring. How thick is the floor?

25. On the average, the human body retains about one-fifth of the food calories taken in. What fraction is not retained?

26. A tank is filled from two pipes, one delivering $2\frac{1}{3}$ gallons per minute, and the other $5\frac{1}{4}$ gallons per minute. At how many gallons per minute is the tank filling?

27. A tank is filled from a pipe at a rate of $5\frac{1}{3}$ gallons per minute. A worker accidentally opens a valve allowing the tank to empty at a rate of $2\frac{1}{5}$ gallons per minute. At what rate is the tank then filling?

28. George read one-half of a book the first day, one-third the next day, and the rest on the third day. What fractional part was read on the third day?

29. An alloy is $\frac{2}{5}$ copper and $\frac{1}{3}$ zinc, and the rest is nickel steel. What fractional part of the alloy is nickel steel?

30. If a mother buys 2 pounds of steak, and junior eats $\frac{7}{8}$ of a pound, father eats $\frac{2}{3}$ of a pound, and baby eats $\frac{3}{8}$ of a pound, how much is left for mother?

Miscellaneous

31. A recipe calls for 2 parts sugar, 5 parts cake mix, and 2 parts water. When mixed together, what fractional part will be sugar?

32. A pipe starts to fill an empty tank at a rate of $16\frac{2}{3}$ gallons per minute. After 21 minutes, a valve is opened that allows the tank to empty at a rate of $22\frac{1}{2}$ gallons per minute. With both valves open, how long will it take to empty the tank?

33. A homemaker buys $2\frac{1}{2}$ pounds of oranges at 32¢ per pound; $3\frac{1}{4}$ pounds of apples at 20¢ per pound; and $5\frac{1}{3}$ pounds of beans at 18¢ per pound. How much did all the produce cost?

34. The outside diameter of a pipe is 1 inch, and the inside diameter is $\frac{3}{4}$ inch. How thick is the wall of the pipe?

35. A $1\frac{5}{16}$-inch hole is punched in the center of a square plate of metal 2 inches on each side. What is length a in the diagram?

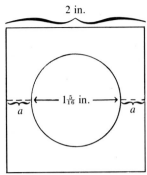

36. An automatic drill takes $4\frac{3}{32}$ seconds to drill a hole in a metal plate $\frac{1}{8}$ inch thick, and it takes $1\frac{1}{15}$ seconds to reset between drillings. How long will it take to drill 64 holes?
(*Warning:* How many spaces are between the holes?)

Name: _____

Class: _____

Unit 4 Test Express as improper fractions.

1. $7\frac{5}{9} =$

2. $13\frac{3}{8} =$

Express as mixed numbers.

3. $\frac{8}{3} =$

4. $\frac{384}{27} =$

Find the products (lowest terms).

5. $3\frac{3}{5} \times 3\frac{3}{4} =$

6. $10\frac{1}{2} \times 2\frac{2}{3} =$

7. $2\frac{2}{5} \times 2\frac{1}{2} \times 2\frac{1}{3} =$

Find the quotients (lowest terms).

8. $2\frac{2}{3} \div 5\frac{1}{3} =$

9. $4\frac{1}{2} \div 3\frac{3}{8} =$

10. $2\frac{2}{15} \div 1\frac{13}{35} =$

Find the sums (lowest terms).

11. $7\frac{7}{8}$
    $+5\frac{5}{8}$

12. $3\frac{2}{3}$
    $+5\frac{1}{2}$

13. $4\frac{5}{24}$
    $+3\frac{7}{18}$

14. $8\frac{5}{12}$
    $2\frac{7}{18}$
    $+6\frac{19}{24}$

Find the differences (lowest terms).

15. $5\frac{7}{9}$
    $-3\frac{4}{9}$

16. $5\frac{3}{8}$
    $-4\frac{3}{4}$

17. $43\frac{5}{36}$
    $-36\frac{7}{24}$

Word problems

18. At a certain speed a car is found to get $17\frac{1}{2}$ miles per gallon of gasoline. How may gallons would be required to drive 350 miles?

19. If an item costs $\$2\frac{1}{2}$ per pound, how much would $1\frac{3}{5}$ pounds cost?

20. A housewife bought $3\frac{1}{2}$ pounds of flour, $5\frac{1}{3}$ pounds of apples, and $7\frac{3}{4}$ pounds of potatoes. What is the total weight she must carry home?

Section 5-1 Decimal Numbers

A **decimal number** is a number that can be expressed as a common fraction whose denominator is equal to a power of 10. The numbers are also expressed in decimal form.

$$\begin{array}{cc} \textit{Decimal form} & \textit{Common-fraction form} \\ .3 & = \frac{3}{10} \\ .29 & = \frac{29}{100} \\ 5.68 & = \frac{568}{100} \\ .764 & = \frac{764}{1000} \end{array}$$

In each case the **number of digits to the right of the decimal point is the same as the number of zeros in the denominator.**

The place value associated with each digit becomes clear when the common fraction is written in expanded form.

$$53.869 = \frac{53,869}{1,000}$$

$$= \frac{50,000 + 3,000 + 800 + 60 + 9}{1,000}$$

$$= \frac{50,000}{1,000} + \frac{3,000}{1,000} + \frac{800}{1,000} + \frac{60}{1,000} + \frac{9}{1,000}$$

$$= 50 \quad + \quad 3 \quad + \quad \frac{8}{10} \quad + \quad \frac{6}{100} \quad + \quad \frac{9}{1,000}$$

5 3 . 8 6 9

9 thousandths
6 hundredths
8 tenths
3 ones
5 tens

A larger decimal number is written below, with place values indicated.

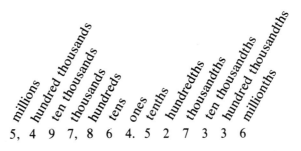

Examples

Decimals written as common fractions and in expanded form:

(a) $.37 = \dfrac{37}{100}$

$= \dfrac{3}{10} + \dfrac{7}{100}$

(b) $.0083 = \dfrac{83}{10,000}$ ← No 0s here

$= \dfrac{8}{1,000} + \dfrac{3}{10,000}$

Note: .0 0 8 3 — no hundredths, no tenths

(c) $358.2934 = \dfrac{3,582,934}{10,000}$

$= 300 + 50 + 8 + \dfrac{2}{10} + \dfrac{9}{100} + \dfrac{3}{1,000} + \dfrac{4}{10,000}$

Common fractions written as decimals and in expanded form:

(d) $\dfrac{261}{1,000} = .261$

$= \dfrac{2}{10} + \dfrac{6}{100} + \dfrac{1}{1,000}$

three zeros in denominator, so count three places from the right and place the decimal: .2 6 1
            3 2 1

(e) $\dfrac{583}{100} = 5.83$

$= 5 + \dfrac{8}{10} + \dfrac{3}{100}$

two zeros, so two places:
5 . 8 3
   2 1

(f) $\dfrac{29}{10,000} = .0029$

$= \dfrac{2}{1,000} + \dfrac{9}{10,000}$

four zeros, so four places:
. 0 0 2 9
 4 3 2 1

Sec. 5-1 Decimal Numbers

Complete each problem below. Completed problems appear on page 158.

Sample Set

(a) $.57 = \dfrac{57}{\Box} = \dfrac{5}{\Box} + \dfrac{7}{\Box}$

(b) $3.852 = \dfrac{\Box}{1{,}000} = \Box + \dfrac{\Box}{10} + \dfrac{\Box}{100} + \dfrac{\Box}{1{,}000}$

(c) $.0063 = \dfrac{\Box}{10{,}000} = \dfrac{6}{\Box} + \dfrac{3}{\Box}$

(d) $\dfrac{537}{100} = \Box\,.\,\Box = \Box + \dfrac{\Box}{10} + \dfrac{\Box}{100}$

(e) $\dfrac{6}{100} = .\Box\Box$

Completed Problems

(a) $\quad .57 = \dfrac{57}{\boxed{100}} = \dfrac{5}{\boxed{10}} + \dfrac{7}{\boxed{100}}$

(b) $\quad 3.852 = \dfrac{\boxed{3,852}}{1,000} = \boxed{3} + \dfrac{\boxed{8}}{10} + \dfrac{\boxed{5}}{100} + \dfrac{\boxed{2}}{1,000}$

(c) $\quad .0063 = \dfrac{\boxed{63}}{10,000} = \dfrac{6}{\boxed{1,000}} + \dfrac{3}{\boxed{10,000}}$

(d) $\quad \dfrac{537}{100} = \boxed{5} \cdot \boxed{37} = \boxed{5} + \dfrac{\boxed{3}}{10} + \dfrac{\boxed{7}}{100}$

(e) $\quad \dfrac{6}{100} = .\boxed{0}\boxed{6}$

Name: _____

Class: _____

Exercises for Sec. 5-1 Write each decimal number as a common fraction and in expanded form.

1. .23
2. .3754
3. 5.78
4. 2.06
5. 46.941
6. 451.3
7. 73.003
8. .0052
9. 700.03
10. .20507

Write each fraction in decimal form and in expanded form.

11. $\dfrac{17}{100}$

12. $\dfrac{632}{1,000}$

13. $\dfrac{632}{100,000}$

14. $\dfrac{52}{10}$

15. $\dfrac{705}{100}$

16. $\dfrac{82,895}{10}$

17. $\dfrac{82,895}{1,000}$

18. $\dfrac{43}{1,000}$

19. $\dfrac{60,203}{100}$

20. $\dfrac{60,203}{100,000}$

Section 5-2 Addition and Subtraction of Decimal Numbers

As in the operations with whole numbers, the commutative and associative laws apply in operations with decimal numbers. This will be demonstrated at the end of the section.

In the problem below, can you see what numbers are being carried or borrowed?

$$\begin{array}{r} \overset{2}{}3.27 \\ 5.38 \\ +1.19 \\ \hline 9.84 \end{array} \qquad \begin{array}{r} \overset{6\,12}{8.6\cancel{7}\cancel{2}} \\ -2.348 \\ \hline 6.324 \end{array}$$

Carried 20 one-hundredths = 2 tenths

Borrowed 1 one-hundredth = 10 one-thousandths

When writing a problem vertically, be sure to line up the decimal points in a straight line. You may find it convenient to place zeros to the right of the decimal point so that each decimal has the same number of places.

$$\begin{array}{r} 13.82 \\ 5.8 \\ +4.356 \\ \hline \end{array} \qquad \begin{array}{r} 13.820 \\ 5.800 \\ +4.356 \\ \hline 23.976 \end{array}$$

$$\begin{array}{r} 5.26 \\ -3.843 \\ \hline \end{array} \qquad \begin{array}{r} 5.260 \\ -3.843 \\ \hline 1.417 \end{array}$$

Complete the problems below. Completed problems appear on page 162.

Sample Set

(a)
$$\begin{array}{r} \square\square \\ 3\,.\,0\;\;8 \\ 2\,.\,5\;\;6 \\ 7\,.\,2\;\;2 \\ +\;5\,.\,9\;\;4 \\ \hline 1\,\square\,.\,8\;\;\square \end{array}$$

(b)
$$\begin{array}{r} .\;7\;\;2\;\;6 \\ 2\,.\,8 \\ +\;6\,.\,3\;\;5 \\ \hline 9\,.\,8\;\;\square\;\;\square \end{array}$$

(c)
$$\begin{array}{r} .\,0\;\;0\;\;7\;\;6 \\ .\,0\;\;6 \\ 8\,.\,2 \\ +\;3\;\;0\,.\,0 \\ \hline \square\,\square\,.\,\square\,\square\,\square\,\square \end{array}$$

(d)
$$\begin{array}{r} 5\,.\,2\;\;8 \\ -\;3\,.\,7\;\;6 \\ \hline 1\,.\,\square\,\;\;2 \end{array}$$

(e)
$$\begin{array}{r} 8\;\;3\,.\,2\;\;5\;\;6 \\ -\;\;\;7\;\;9\,.\,6 \\ \hline \square\,.\,\square\,\square\,\square \end{array}$$

(f)
$$\begin{array}{r} 3\;\;7\,.\,8 \\ -\;2\;\;3\,.\,6\;\;9\;\;7 \\ \hline 1\;\;4\,.\,\square\,\square\,\square \end{array}$$

The Commutative and Associative Laws Used in Addition

$$\begin{array}{r} 5.8 \\ +3.6 \\ \hline 9.4 \end{array}$$

Written horizontally, the problem is:

$5.8 + 3.6 = (5 + .8) + (3 + .6)$ commutative and associative
$ = (5 + 3) + (.8 + .6)$ laws are used to group numbers from each column

$ = 8 + (1 + .4)$ 8 tenths + 6 tenths = 1 + 4 tenths

$ = (8 + 1) + .4$ associative law is used
$ = 9 + .4 = 9.4$ to carry 1

Associative Law Used in Subtraction

$$\begin{array}{r} 5.3 \\ -3.8 \end{array} \rightarrow \begin{array}{r} 5 \;+\; .3 \\ -(3 \;+\; .8) \end{array}$$

$$\rightarrow \begin{array}{r} (4+1) \;+\; .3 \\ -(3 \;\;+\;\; .8) \end{array}$$

$$\rightarrow \begin{array}{r} 4 \;+\; (1 + .3) \\ -(3 \;+\;\;\;\; .8) \end{array}$$ associative law is used to borrow 1

$$\rightarrow \begin{array}{r} 4 \;+\; 1.3 \\ -(3 \;+\; .8) \\ \hline 1 \;+\; .5 = 1.5 \end{array}$$

Completed Problems

(a)
$$\begin{array}{r} \boxed{1}\;\;\boxed{2} \\ 3\,.\,0\;8 \\ 2\,.\,5\;6 \\ 7\,.\,2\;2 \\ +\;5\,.\,9\;4 \\ \hline 1\,\boxed{8}\,.\,8\,\boxed{0} \end{array}$$

(b)
$$\begin{array}{r} .7\;2\;6 \\ 2\,.\,8 \\ +\;6\,.\,3\;5 \\ \hline 9\,.\,8\,\boxed{7}\,\boxed{6} \end{array}$$

(c)
$$\begin{array}{r} .0\;0\;7\;6 \\ .0\;6 \\ 8\,.\,2 \\ +\;3\;0\,.\,0 \\ \hline \boxed{3}\,\boxed{8}\,.\,\boxed{2}\,\boxed{6}\,\boxed{7}\,\boxed{6} \end{array}$$

(d)
$$\begin{array}{r} 5\,.\,2\;8 \\ -\;3\,.\,7\;6 \\ \hline 1\,.\,\boxed{5}\;2 \end{array}$$

(e)
$$\begin{array}{r} 8\;3\,.\,2\;5\;6 \\ -\;7\;9\,.\,6 \\ \hline \boxed{3}\,.\,\boxed{6}\,\boxed{5}\,\boxed{6} \end{array}$$

(f)
$$\begin{array}{r} 3\;7\,.\,8 \\ -\;2\;3\,.\,6\;9\;7 \\ \hline 1\;4\,.\,\boxed{1}\,\boxed{0}\,\boxed{3} \end{array}$$

Name: _____

Class: _____

Exercises for Sec. 5-2 Find the sums.

1. .25
    .06
    .72
    +.89
    ———

2. .7
    .32
    .05
    +.93
    ———

3. .0021
    .0095
    .0087
    +.0044
    ————

4. 2.0
    5.3
    8.9
    +7.5
    ———

5. 6.2
    2.5
    5.4
    +7.8
    ———

6. .629
    .06
    .052
    +.73
    ———

7. .82
    3.45
    4.68
    +9.21
    ————

8. 3.803
    .765
    2.72
    +5.325
    —————

9. 7.569
    2.93
    4.372
    +5.86
    —————

10. 32.5
    56.7
    22.8
    +83.7
    ————

11. 42.75 + 39.48 + 24.61 + 10.32 =

163

12. 5.43 + .856 + 7.429 + 5.68 =

13. 175.2 + 283.7 + 161.8 + 854.2 =

14. 5.32 + .0271 + .0946 + .0756 =

15. .0038 + .0271 + .0946 + .0756 =

16. 5.2 + .057 + 3.7 + 2.06 =

17. 4.737 + 2.05 + .982 + 1.07 =

18. 253 + 5.29 + 3.054 + .869 =

19. 12.8 + 5.07 + 6.309 + 1.64 =

20. 384.06 + 742.83 + 15.76 + 493.27 =

Find the differences.

21. 3.4
    −2.7

22. 8.3
    −5.8

23. 97.2
    −38.6

24. 54.7
    − 7.8

25. 3.06
    −2.9

26. .0207
    −.0186

27. 5.62
    −2.09

28. 7.98
    −6.99

29. 5.06
    −4.97

30. 8.006
    − .007

31. 4.8
    −2.94

32. 5.070
    − .608

33. 7.8 − 4.632 =

34. 4.87 − 3.9 =

35. $84.06 - 83.97 =$

36. $4.006 - 3.94 =$

37. $16.73 - 10.08 =$

38. $2.06 - .0072 =$

39. $58.77 - 48.78 =$

40. $3 - 2.997 =$

Section 5-3 Multiplication of Numbers with Decimal Part

Two decimal numbers are multiplied below.

$$\underbrace{5.32}_{\text{2 places}} \times \underbrace{4.3}_{\text{1 place}} = \underbrace{22.876}_{\text{3 places}}$$

Notice that the number of decimal places in the product is the sum of the number of decimal places in the factors. The reason for this is easily seen when the numbers are written as common fractions.

$$\underbrace{\frac{532}{100}}_{\text{2 zeros}} \times \underbrace{\frac{43}{10}}_{\text{1 zero}} = \underbrace{\frac{22,876}{1,000}}_{\text{3 zeros}}$$

Adding the number of decimal places is equivalent to adding the number of zeros when powers of 10 are multiplied in the denominators.

Another example:

$$\underbrace{.32}_{\text{2 places}} \times \underbrace{.026}_{\text{3 places}} = \underbrace{.00832}_{\text{5 places (from the right)}}$$

$$\underbrace{\frac{32}{100}}_{\text{2 zeros}} \times \underbrace{\frac{26}{1,000}}_{\text{3 zeros}} = \underbrace{\frac{832}{100,000}}_{\text{5 zeros}}$$

Place the decimal point in each product below. Completed products appear on page 168.

Sample Set

Except for placement zeros and decimal points, the digits below each problem are correct.

(a) 8.74
   × .26
   ─────
   22724

(b) 87.4
   × 2.6
   ─────
   22724

(c) 8.74
   × 2.6
   ─────
   22724

(d) .874
   × .26
   ─────
   22724

(e) 8740
   × .026
   ─────
   227240

(f) .874
   ×.026
   ─────
   22724

(g) .0874
   × .026
   ─────
   22724

(h) .0874
   × 2.6
   ─────
   22724

(i) .874
   ×2600
   ─────
   2272400

Completed Products

(a) 2.2724	(b) 227.24	(c) 22.724
(d) .22724	(e) 227.240	(f) .022724
(g) .0022724	(h) .22724	(i) 2,272.400

Exercises for Sec. 5-3 Find the products.

1. 3.2
   ×2.8

2. .32
   ×.28

3. 3.2
   ×.28

4. .032
   × 2.8

5. .032
   ×.028

6. 843
   × .7

7. 9.44
   ×.006

8. .482
   × .7

9. 1.58
   × .08

10. 3.34
    × .6

11. 5.81
    × 3

12. 81.9
    × .8

13. .736
    × .05

14. 22.7
    ×.005

15. .649
    ×.007

16. 853
    ×.004

17. 7.28
    × .32

18. 59.3
    × .21

19. .647
   × 38

20. 20.3
   ×.084

21. 75.1
   × 7.2

22. .0689
   × .43

23. 343
   ×.87

24. .0956
   × 5.3

25. .782
   × 4.7

26. 29.4
   ×.062

27. 3.36
   ×.028

28. 4.36
   ×.084

29. .0529
   × .034

30. .0496
   ×.0026

Section 5-4 Rounding Decimal Numbers

Examples of rounding are given below.

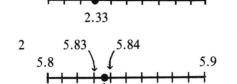

1. 2.33 is closer to 2.3 than to 2.4; rounded to the nearest 10th, 2.33 is, therefore, 2.3

2. 5.837 is closer to 5.8 than to 5.9; rounded to the nearest 10th, 5.837 is 5.8

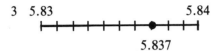

3. 5.837 is closer to 5.84 than to 5.83; rounded to the nearest 100th, 5.837 is 5.84

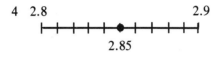

4. 2.85 is halfway between 2.8 and 2.9; to the "nearest" 10th, it is customary to round 2.85 "up" to 2.9

When rounding a number to the nearest 10th, observe the digit in the hundredths' place. If it is equal to or greater than 5, round up. If it is less than 5; round down. The same rule holds when rounding to any given place.

To the nearest 10th,

37.86	rounds up to	37.9
4.75	rounds up to	4.8
7.96	rounds up to	8.0 (not 8)

↗ zero indicates rounding to the nearest 10th

| 87.63 | rounds down to | 87.6 |
| 36.04 | rounds down to | 36.0 |

The numbers below are rounded to the nearest 1,000th, 100th, and 10th:

Number	Nearest 1,000th	Nearest 100th	Nearest 10th	
75.3476	75.348	75.35	75.3	← to get this, round 75.3476, not 75.35
6.7962	6.796	6.80	6.8	
8.9996	9.000	9.00	9.0	

Decimal Arithmetic

A number may also be rounded to the nearest 1s, 10s, or 100s place:

Number	Nearest 1	Nearest 10	Nearest 100	
5,736.8	5,737	5,740	5,700	
8,799.603	8,800	8,800	8,800	
38.3	38	40	0	← 38.3 is closer to 0 than to 100
447.29	447	450	400	← to get this, round 447.29, not 450

Complete the problems below. Completed problems appear on page 173.

Sample Set

Round each number below to the nearest 1,000th, 100th, and 10th.

	Number	Nearest 1,000th	Nearest 100th	Nearest 10th
(a)	5.3728	5 . 3 7 ☐	5 . 3 ☐	5 . ☐
(b)	.83252	. 8 3 ☐	. 8 ☐	. ☐
(c)	1.4003	1 . 4 0 ☐	1 . 4 ☐	1 . ☐
(d)	.0328	. 0 3 ☐	. 0 ☐	. ☐
(e)	6.9999	☐ . ☐☐☐	☐ . ☐☐	☐ . ☐

Round each number below to the nearest 1, 10 and 100.

	Number	Nearest 1	Nearest 10	Nearest 100
(f)	584.7	5 8 ☐	5 ☐ 0	☐ 0 0
(g)	793.8	7 9 ☐	☐ ☐ 0	☐ ☐ 0
(h)	5,287.5	5 , 2 8 ☐	5 , 2 ☐ ☐	5 , ☐ ☐ ☐
(i)	38.2	3 ☐	☐ 0	☐

Completed Problems

	Number	Nearest 1,000th	Nearest 100th	Nearest 10th
(a)	5.3728	5.373	5.37	5.4
(b)	.83252	.833	.83	.8
(c)	1.4003	1.400	1.40	1.4
(d)	.0328	.033	.03	.0
(e)	6.9999	7.000	7.00	7.0

	Number	Nearest 1	Nearest 10	Nearest 100
(f)	584.7	585	580	600
(g)	793.8	794	790	800
(h)	5,287.5	5,288	5,290	5,300
(i)	38.2	38	40	0

Name: _____

Class: _____

Exercises for Sec. 5-4 Round each number to the nearest 1,000th, 100th, and 10th.

	Number	Nearest 1,000th	Nearest 100th	Nearest 10th
1.	8.4789			
2.	.83653			
3.	37.98229			
4.	7.21083			
5.	4.827391			
6.	.97360			
7.	.63092			
8.	5.000			
9.	.00029			
10.	3.9995			
11.	62.70135			
12.	14.39251			
13.	.402991			
14.	1.39025			
15.	37.943251			

Round each number to the nearest 10,000th.

16. 7.084763   17. 5.927549
18. .682981   19. 1.00002
20. 5.03997

Round each number to the nearest 1, 10, and 100.

	Number	Nearest 1	Nearest 10	Nearest 100
21.	375.7			
22.	596.2			
23.	93.52			
24.	809.9			
25.	983.48			
26.	5,267.2			
27.	6,750.1			
28.	7,399.7			
29.	8,999.6			
30.	3.2			

Section 5-5 Division of Decimal Numbers

In a division problem, the quotient is not changed when the divisor and dividend are multiplied by the same power of 10.

$$7 \overline{)28} \qquad 70 \overline{)280} \qquad 700 \overline{)2800}$$
$$\phantom{7 \overline{)28}} \qquad \text{divisor and} \qquad \text{divisor and}$$
$$\phantom{7 \overline{)28}} \qquad \text{dividend} \times 10 \qquad \text{dividend} \times 100$$

The quotients are 4, 4, and 4.

This also holds true when decimal numbers are divided. Multiplying the divisor and dividend by a power of 10 justifies the familiar "shifting" of the decimal point.

quotient $Q \times 1 = Q$

1. $.6 \overline{)4.3.2}$ gives 7.2
divisor and dividend × 10

$$4.32 \div .6 = \frac{4.32}{.6} \times \frac{10}{10} = \frac{43.2}{6} = 4.32 \div 6$$

2. $.02 \overline{)_.05.36}$ gives 2.68
divisor and dividend × 100

$$.0536 \div .02 = \frac{.0536}{.02} \times \frac{100}{100} = \frac{5.36}{2}$$
$$= 5.36 \div 2$$

3. $.003 \overline{)6.279.}$ gives $2\,093.$
divisor and dividend × 1,000

$$6.279 \div .003 = \frac{6.279}{.003} \times \frac{1{,}000}{1{,}000} = \frac{6{,}279}{3}$$
$$= 6{,}279 \div 3$$

The decimal point is shifted to the right until the divisor becomes a whole number. The decimal point in the dividend must be shifted the same number of places.

Looking at the problems above, can you see why you would want to shift the decimal point? This is done only to show us where the decimal point is placed in the quotient. **After the decimal point is shifted, the decimal point in the quotient will be directly above the decimal point in the dividend.**

Repeating Decimal   A decimal is called a **repeating decimal** if a digit, or set of digits, repeats indefinitely.

.033333 · · ·   3 repeats
41.5151515 · · ·   15 repeats

A special case of repeating decimals is the decimal in which the digit 0 repeats.

3.750000 · · ·
.796000 · · ·
.062000 · · ·

Decimal Arithmetic

Repeating zeros occur in quotients when the division terminates with remainder zero.

Examples | Division terminating with remainder zero:

(a)   .3 ÷ .08

$$.08\overline{\smash{)}.30.00}$$
```
        3.75
.08 ) .30.00
       24
        6 0
        5 6
          40
          40
           0
```
0s may be placed after the 3

(b)   1.0348 ÷ 1.3
```
           .796
1.3 ) 1.0.348
         9 1
         1 24
         1 17
            78
            78
             0
```

(c)   .01302 ÷ .21
```
          .062
.21 ) .01.302
        1 26
          42
          42
           0
```

Division not terminating with remainder zero:

(d)   .1 ÷ 3

since the divisor 3 is a whole number, the decimal will not be shifted
```
       .0333 · · ·
3 ) .1000
     9
     10  ← a repeating remain-
      9   /  der tells us that a
     10  /   repeating decimal
              will occur
```

Rounded to the nearest 1,000th, the quotient is .033.

(e) $137 \div 3.3$

the whole number 137 can be written as a decimal: 137.000

$$\begin{array}{r} 41.51515\cdots \\ 3.3\overline{\smash{\big)}\,137.0\,00} \\ \underline{132} \\ 5\,0 \leftarrow \text{repeating} \\ \underline{3\,3} \text{remainder} \\ 1\,7\,0 \\ \underline{1\,6\,5} \\ 50 \end{array}$$

Rounded to the nearest 100th, the quotient is 41.52. To the nearest 1,000th, the quotient is 41.515.

In the problems below, place the decimal point in each quotient. If necessary, put in placement zeros. Completed quotients appear on page 180.

Sample Set

$$32\overline{\smash{\big)}\,928} \qquad 29$$
$$\underline{64}$$
$$288$$
$$\underline{288}$$
$$0$$

(a) $32\overline{\smash{\big)}\,9.28}$ — quotient 29

(b) $32\overline{\smash{\big)}\,.928}$ — quotient 29

(c) $32\overline{\smash{\big)}\,.0928}$ — quotient 29

(d) $3.2\overline{\smash{\big)}\,9.28}$ — quotient 29

(e) $3.2\overline{\smash{\big)}\,92.8}$ — quotient 2 9

(f) $.32\overline{\smash{\big)}\,9.28}$ — quotient 29

(g) $.32\overline{\smash{\big)}\,.928}$ — quotient 29

(h) $.032\overline{\smash{\big)}\,92.8}$ — quotient 2 9

(i) $.032\overline{\smash{\big)}\,.928}$ — quotient 29

(j) $.032\overline{\smash{\big)}\,.00928}$ — quotient 29

Completed Quotients

(a) .29	(b) .029	(c) .0029	(d) 2.9
(e) 29	(f) 29	(g) 2.9	(h) 2900
(i) 29	(j) .29		

Name: _____

Class: _____

Exercises for Sec. 5-5 Find the quotients. If the division does not terminate with remainder zero, round the quotient to the nearest 1,000th.

1. 5.3) 4.611   2. 5.3) 461.1   3. .53) 4.611

4. .53) 461.1   5. .53) .04611   6. 53) .04611

7. 2) .5163   8. .7) .0784   9. 9) 7.384

10. .05) 7.32   11. .8) .0397   12. .06) 52.7

13. $7\overline{).0032}$ 14. $.003\overline{).0572}$ 15. $.3\overline{)377}$

16. $1.8\overline{).0871}$ 17. $.24\overline{).0576}$ 18. $2.8\overline{)15.96}$

19. $32\overline{).087}$ 20. $.20\overline{)385.6}$ 21. $.13\overline{)54.6}$

22. $1.6\overline{).0832}$ 23. $.43\overline{).3741}$ 24. $3.6\overline{).00648}$

Sec. 5-5 Division of Decimal Numbers

25. .009) 2.6   26. .47) 13.301   27. 73) .05928

28. .015) .00629   29. .015) .629   30. .15) 629

Section 5-6 Writing Fractions as Equivalent Decimals

The decimal representation of a common fraction is found by dividing the numerator by the denominator.

Examples with Common Fractions

(a) $\frac{2}{5} = .4$

$$\begin{array}{r} .4 \\ 5\overline{)2.0} \\ \underline{2\ 0} \\ 0 \end{array}$$

(b) $\frac{3}{8} = .375$

$$\begin{array}{r} .375 \\ 8\overline{)3.000} \\ \underline{2\ 4} \\ 60 \\ \underline{56} \\ 40 \\ \underline{40} \\ 0 \end{array}$$

(c) Rounded to the nearest 1,000th, $\frac{7}{6} = 1.167$

$$\begin{array}{r} 1.166 \\ 6\overline{)7.000} \\ \underline{6} \\ 1\ 0 \\ \underline{6} \\ 40 \longleftarrow \text{repeats} \\ \underline{36} \\ 40 \\ \underline{36} \\ 4 \end{array}$$

Examples with Mixed Numbers

(a) $4\frac{3}{4} = 4.75$

$$\begin{array}{r} .75 \\ 4\overline{)3.00} \\ \underline{2\ 8} \\ 20 \\ \underline{20} \\ 0 \end{array}$$

(b) Rounded to the nearest 1,000th, $8\frac{5}{9} = 8.556$

$$\begin{array}{r} .555 \\ 9\overline{)5.000} \\ \underline{4\ 5} \\ 50 \longleftarrow \text{repeats} \\ \underline{45} \\ 50 \\ \underline{45} \\ 5 \end{array}$$

(c) Rounded to the nearest 1,000th,
$12\frac{6}{7} = 12.857$

```
        .8571
     7)6.0000
       5 6
       ---
        40
        35
        --
        50
        49
        --
        10
         7
        --
         3
```

Since any whole number can be written as a decimal, the process of converting a fraction to a decimal is a special case of decimal division.

Rational and Irrational Numbers

A number that can be represented in the form a/b, where a and b are whole numbers, and $b \neq 0$, is called a **rational number**.

Examples
$\frac{3}{5}$ **numbers represented as proper fractions**
$1\frac{2}{3} = \frac{5}{3}$ **mixed numbers**
$7 = \frac{7}{1}$ **whole numbers**
$3.25 = \frac{325}{100}$ **decimal numbers**

Numbers that are not rational numbers are called **irrational numbers**. Looking at the above examples, one might conclude that there are no irrational numbers. This is not true. There is an endless number of irrationals.

An example of an irrational number is $\sqrt{2}$, a number which, when multiplied by itself, gives the product 2 ($\sqrt{2} \times \sqrt{2} = 2$).

Another example is π, the quotient when the circumference of a circle is divided by the diameter. The rational numbers $\frac{22}{7}$ and 3.14 which are often used to represent π are only approximations.

It is a fact that when the numerator of a common fraction is divided by the denominator, the quotient is a repeating decimal (possibly with repeating zeros). **Rational numbers are therefore those numbers which can be represented by repeating decimals.**

The digits in the representation of an irrational number do not repeat in any pattern.

When decimal numbers are divided, the quotient can be represented as a common fraction. For example,

$$.29 \div .7 = \frac{.29}{.7} = \frac{.29}{.7} \times \frac{100}{100} = \frac{29}{70}$$

Since common fractions are equivalent to repeating decimals, the quotient obtained when decimals are divided must be a repeating decimal (possibly with repeating zeros).

Name: _____

Class: _____

Exercises for Sec. 5-6 Express each fraction as a decimal. If the division does not terminate with remainder zero, round the decimal to the nearest 1,000th.

1. $\frac{1}{2} =$ 2. $\frac{3}{4} =$ 3. $\frac{3}{5} =$

4. $\frac{4}{5} =$ 5. $\frac{2}{3} =$ 6. $\frac{2}{9} =$

7. $\frac{4}{9} =$ 8. $\frac{7}{9} =$ 9. $\frac{1}{8} =$

10. $\frac{3}{8} =$ 11. $\frac{9}{8} =$ 12. $\frac{11}{8} =$

13. $\frac{1}{6} =$ 14. $\frac{2}{7} =$ 15. $\frac{18}{7} =$

16. $\frac{5}{6} =$ 17. $\frac{5}{12} =$ 18. $\frac{26}{15} =$

19. $\frac{7}{20} =$ 20. $\frac{5}{25} =$ 21. $\frac{43}{18} =$

22. $\frac{5}{24} =$ 23. $\frac{4}{29} =$ 24. $\frac{11}{23} =$

25. $\frac{95}{32} =$ 26. $\frac{121}{50} =$ 27. $\frac{7}{251} =$

28. $\frac{5}{173} =$ 29. $\frac{3}{1523} =$ 30. $\frac{7}{1253} =$

Sec. 5-6 Writing Fractions as Equivalent Decimals

31. $5\frac{3}{4} =$ 32. $6\frac{7}{8} =$ 33. $2\frac{2}{3} =$

34. $1\frac{5}{6} =$ 35. $4\frac{5}{8} =$ 36. $7\frac{2}{7} =$

37. $3\frac{5}{13} =$ 38. $9\frac{6}{11} =$ 39. $7\frac{5}{24} =$

40. $8\frac{5}{29} =$

Name: _____

Class: _____

Applications (Round nonterminating division to the nearest 1,000th.)

1. If a plane traveled 2,200 miles in 4.7 hours, what was its average speed?

2. A student bought a new book for $10.95. At the end of the semester, he sold the book for $3.50. How much did he lose?

3. A man earned $7.85 per hour. How much did he earn if he worked 32.5 hours?

4. If gasoline costs $.675 per gallon, how much will 18.3 gallons cost?

5. A woman's gross pay was $475.50 for a week. Her employer deducted $83.27 for taxes, $13.50 for insurance, and $53.29 for her pension. How much was her take-home pay?

6. One cubic foot of water weighs 62.5 pounds, and 1 cubic foot of ice weighs 57.5 pounds. How much heavier is a cubic foot of water than a cubic foot of ice?

7. Gold weighs approximately 1,200 pounds per cubic foot, and lead weighs approximately 710 pounds per cubic foot. Gold is how many times heavier than lead?

8. A dealer makes $4.59 profit on each coat he sells. How much profit will be made if he sells 50 coats?

9. Water is pumped into a tank at a rate of 32.7 gallons per minute. If a valve at the bottom of the tank is opened and water runs out at 7.5 gallons per minute, at what rate will the tank fill?

10. How far will a car traveling 63.5 miles per hour go in 10.7 hours?

11. A shopper found apples listed at $.19 per pound. He put 10 apples in a bag and found they weighed 7.3 pounds. How much should the apples cost?

12. If carpet is selling for $12.75 per square yard, how much will 24 square yards cost?

13. How long will it take a car to travel 358.8 miles if the car travels 55.2 miles per hour?

14. A pump delivers 17.3 gallons per second. How many gallons will it deliver in 60 seconds?

15. If a pump delivers 17.3 gallons per second, how many seconds will it take to fill a 1,000-gallon tank?

16. A tank is filled from three pipes delivering 75.2 gallons per minute, 57.8 gallons per minute, and 29.5 gallons per minute. At what rate are the three pipes together filling the tank?

17. An investor has $14,625 to invest in stocks. If each share costs $32.50, how many shares can be bought?

18. A company's net income was $1,762,062.50. If the company had 582,500 outstanding shares of stock, what was the net earning per share?

19. If rolled copper weighs 548.72 pounds per cubic foot, how much will 12.3 cubic feet weigh?

20. If a 5-pound bag of sugar costs $.89, and a 25-pound bag costs $3.97, how much would you save by buying the 25-pound bag instead of five 5-pound bags?

21. A football player ran for a total of 1,273 yards during a 14-game season. How many yards did he average per game?

22. A farmer harvested 6,527 bushels of corn from a 62-acre field. How many bushels did he average per acre?

23. A man bought 55 shares of stock at $17.625 per share and sold the stock for $11.375 per share. How much did he lose?

24. A wholesaler sold 27 sport coats for $28.75 each and 45 pairs of pants for $8.25 per pair. How much was the entire order?

25. A metal plate .235 inch thick is painted on both sides. If the paint is .073 inch thick, how thick will the painted plate be?

26. A contractor hired seven carpenters at $13.50 per hour, two plumbers at $14.75 per hour, and three electricians at $14.30 per hour. What total wage did the contractor pay per hour?

27. Number 9 gauge steel plate is .15625 inch thick, and no. 5 gauge steel plate is .21875 inch thick. If the walls of a boiler are made with one layer of no. 9 plate and two layers of no. 5 plate, how thick will the walls be?

28. What is the outside diameter D of a pipe if the inside diameter is 1.75 inches and the thickness of the metal is .1875 inch.

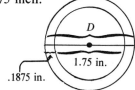

29. A pipe 22.2 inches long is wound with wire that has a cross-section diameter of .148 inch. How many times must the wire be wound around the pipe in order to cover the entire pipe?

30. One cubic foot of water weighs 62.5 pounds, and 1 cubic foot of ice weighs 57.5 pounds. If 1 cubic foot of ice melts, how much volume will it occupy?

Name: _____

Class: _____

Unit 5 Test 1. Write .056 as a common fraction and in expanded form.

2. Write 73.82 as a mixed number and in expanded form.

3. Write $40 + 2 + \frac{5}{10} + \frac{3}{100}$ as a decimal and as a mixed number.

Compute each answer.

4.   5.2
     +4.8

5.   3.5
      5.6
     +4.5

6.   35.72
       4.9
     + 4.386

7.   7.85
     −4.67

8.   527.34
     − 38.7

9.   4.6
     −3.74

10.   .6
     × 5

11.   5.1
     ×4.7

12.   8.5
     ×.06

13.   .52
     ×6.7

14. $.9 \overline{)5.4}$

15. $.7 \overline{)3.71}$

16. $3.5 \overline{) .735}$ 17. $.67 \overline{) .02144}$

18. A builder bought 15.2 tons of gravel and used 7.8 tons. How much did he have left?

19. A motorist drove 196.7 miles in 3.5 hours. What was his average speed?

20. A consumer purchased 3.2 pounds of meat at $1.65 per pound, and 5.3 pounds of produce for $.23 per pound. What was the cost of the entire purchase?

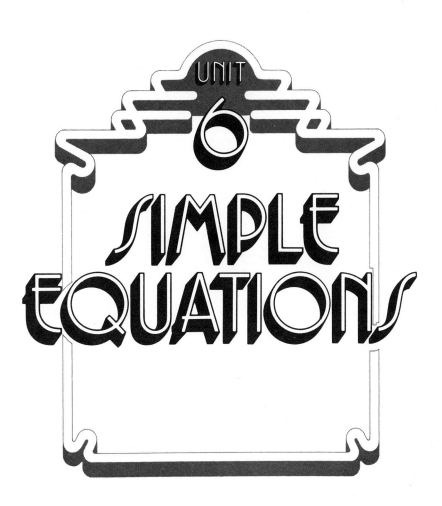

Section 6-1 Equations of the Form $ax = c$ and $ax/b = c$

Letters are often used in mathematics to represent numbers. When x represents a number, $2x$ represents twice x, or 2 times x. If $x = 5$, then $2x = 2(5) = 10$.

If $x = 4$, then
$$2x = 2(4) = 8$$
$$3x = 3(4) = 12$$
$$15x = 15(4) = 60$$

The equality $5x = 30$ is called an **equation**. In words, the equation states: "5 times x equals 30." What must x be? Clearly $x = 6$. The number $x = 6$ is called the **solution** to the equation.

Can you guess the solution to each of the equations below?

1. $7x = 56$
2. $5x = 20$
3. $9x = 72$

Answers
1. $x = 8$ since $7(8) = 56$
2. $x = 4$ since $5(4) = 20$
3. $x = 8$ since $9(8) = 72$

It is not always easy to guess the solution. Can you guess the solution to the equation below?

$$7x = 32$$

The following law of logic can aid you in finding the solution.

> If $a = b$ and $c \neq 0$, then $\dfrac{a}{c} = \dfrac{b}{c}$
>
> ↑ not equal

In words, when equal quantities are divided by the same number, the quotients will be equal. Recall that you cannot divide by zero.

In order to solve the equation $7x = 32$, divide each side of the equation by 7.

$$\frac{7x}{7} = \frac{32}{7}$$

$$x = 4\frac{4}{7}$$

Notice that x appears alone on one side of the equation. "Getting x alone" may be viewed as the goal when solving an equation.

The answer can be checked by substituting the answer back into the equation. If the product obtained is 32, the answer is correct.

$$\text{Check:} \quad 7\left(4\tfrac{4}{7}\right) \stackrel{?}{=} 32$$

$$7\left(\frac{32}{7}\right) \stackrel{?}{=} 32$$

$$32 = 32 \quad \text{true, so the answer was correct}$$

When an equation is in balance, the quantities on each side are equal. This is indicated by the picture of the beam balance on the left:

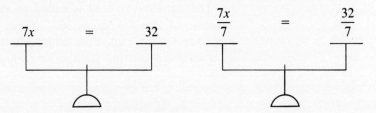

If the quantity on the left side of the balance is divided by 7, the left side will go up. In order to keep the system in balance, the quantity on the right side must also be divided by 7.

Generally, what you do to one side of a balanced equation, you must do to the other in order to maintain the balance.

A second law of logic can also be used when solving equations.

$$\text{If } a = b \quad \text{then } ca = cb$$

In words, when equal quantities are multiplied by the same number, the products are equal. You may find this law easier to use when fractions appear in an equation.

Examples

1. $5x = \dfrac{2}{3}$ $\qquad$ Check: $\cancel{5}\left(\dfrac{2}{\underset{3}{\cancel{15}}}\right) \stackrel{?}{=} \dfrac{2}{3}$

$\dfrac{1}{\cancel{5}}(\cancel{5}x) = \dfrac{1}{5}\left(\dfrac{2}{3}\right)$ $\qquad\qquad \dfrac{2}{3} = \dfrac{2}{3}$

$x = \dfrac{2}{15}$

2. $\dfrac{2}{3}x = 5$ $\qquad$ Check: $\dfrac{2}{\cancel{3}}\left(\dfrac{\overset{5}{\cancel{15}}}{\cancel{2}}\right) \stackrel{?}{=} 5$

$\dfrac{\cancel{3}}{\cancel{2}}\left(\dfrac{\cancel{2}}{\cancel{3}}x\right) = \dfrac{3}{2}\left(\dfrac{5}{1}\right)$ $\qquad\qquad 5 = 5$

$x = \dfrac{15}{2} = 7\dfrac{1}{2}$

3 In this example, x is divided by 9. (*Note:* $x/9 = \frac{1}{9}x$)

$$\frac{x}{9} = 4 \qquad \text{Check:} \quad \frac{36}{9} = 4$$

$$\cancel{9}\left(\frac{x}{\cancel{9}}\right) = 9(4)$$

$$x = 36$$

Examples continue on page 204.

Complete the problems below. Completed problems appear on page 204.

Sample Set

Equation: *Solution:*

(a) $9x = 15$ $\dfrac{9x}{\square} = \dfrac{15}{\square} = \dfrac{\square}{\square}$

$x = \square$

(b) $3x = \dfrac{2}{5}$ $\dfrac{\square}{\square}(3x) = \dfrac{\square}{\square}\left(\dfrac{2}{5}\right)$

$x = \square$

(c) $\dfrac{1}{2}x = 7$ $\dfrac{\square}{\square}\left(\dfrac{1}{2}x\right) = \dfrac{\square}{\square}\left(\dfrac{7}{1}\right)$

$x = \square$

(d) $\dfrac{5}{4}x = \dfrac{2}{10}$ $\dfrac{\square}{\square}\left(\dfrac{5}{4}x\right) = \dfrac{\square}{\square}\left(\dfrac{2}{10}\right) = \dfrac{\square}{\square}$

$x = \square$

(e) $\dfrac{3x}{8} = \dfrac{9}{4}$ $\dfrac{\square}{\square}\left(\dfrac{3x}{8}\right) = \dfrac{\square}{\square}\left(\dfrac{9}{4}\right) = \dfrac{\square}{\square}$

$x = \square$

204 Simple Equations

4 In this example, x has been multiplied by 2. That product was then divided by 5. (*Note:* $2x/5 = \frac{2}{5}x$)

$$\frac{2x}{5} = 7 \qquad\qquad \text{Check: } \frac{\cancel{2}^1}{\cancel{5}_1}\left(\frac{\cancel{35}^7}{\cancel{2}_1}\right) \stackrel{?}{=} 7$$

$$\frac{5}{2}\left(\frac{2x}{5}\right) = \frac{5}{2}\left(\frac{7}{1}\right) \qquad\qquad 7 = 7$$

$$x = \frac{35}{2} = 17\frac{1}{2}$$

Completed Problems

Equation: Solution:

(a) $9x = 15$ $\quad \dfrac{9x}{\boxed{9}} = \dfrac{15}{\boxed{9}} = \dfrac{\boxed{5}}{\boxed{3}}$

$\qquad\qquad x = \boxed{1\,2/3}$

(b) $3x = \dfrac{2}{5}$ $\quad \dfrac{\boxed{1}}{\boxed{3}}(3x) = \dfrac{\boxed{1}}{\boxed{3}}\left(\dfrac{2}{5}\right)$

$\qquad\qquad x = \boxed{2/15}$

(c) $\dfrac{1}{2}x = 7$ $\quad \dfrac{\boxed{2}}{\boxed{1}}\left(\dfrac{1}{2}x\right) = \dfrac{\boxed{2}}{\boxed{1}}\left(\dfrac{7}{1}\right)$

$\qquad\qquad x = \boxed{14}$

(d) $\dfrac{5}{4}x = \dfrac{2}{10}$ $\quad \dfrac{\cancel{\boxed{4}}}{\boxed{5}}\left(\dfrac{5}{4}x\right) = \dfrac{\cancel{\boxed{4}}}{5}\left(\dfrac{\cancel{2}^2}{\cancel{10}_5}\right) = \dfrac{\boxed{4}}{\boxed{25}}$

$\qquad\qquad x = \boxed{4/25}$

(e) $\dfrac{3x}{8} = \dfrac{9}{4}$ $\quad \dfrac{\cancel{\boxed{8}}}{\boxed{3}}\left(\dfrac{3x}{8}\right) = \dfrac{\cancel{\boxed{8}}^2}{\cancel{\boxed{3}}_1}\left(\dfrac{\cancel{9}^3}{\cancel{4}_1}\right) = \dfrac{\boxed{6}}{\boxed{1}}$

$\qquad\qquad x = \boxed{6}$

Name: _____

Class: _____

Exercises for Sec. 6-1   Solve the equations. Check your answers.

1. $3x = 12$

2. $5x = 35$

3. $7x = 56$

4. $12x = 84$

5. $11x = 121$

6. $24x = 332$

7. $4x = 9$

8. $5x = 3$

9. $9x = \dfrac{3}{5}$

10. $8x = 15$

11. $20x = 45$

12. $5x = \dfrac{15}{4}$

13. $16x = \dfrac{12}{5}$ 14. $25x = \dfrac{45}{2}$

15. $28x = 42$ 16. $\dfrac{2}{3}x = 7$

17. $\dfrac{x}{3} = 4$ 18. $\dfrac{x}{5} = 7$

19. $\dfrac{x}{3} = \dfrac{2}{5}$ 20. $\dfrac{x}{4} = \dfrac{3}{7}$

21. $\dfrac{x}{8} = \dfrac{5}{12}$ 22. $\dfrac{x}{6} = \dfrac{7}{9}$

23. $\dfrac{2x}{5} = 3$ 24. $\dfrac{3x}{7} = 5$

Sec. 6-1 Equations of the Form $ax = c$ and $ax/b = c$

25. $\dfrac{5}{8}x = 10$

26. $\dfrac{8}{9}x = 12$

27. $\dfrac{3}{4}x = \dfrac{27}{36}$

28. $\dfrac{4}{15}x = \dfrac{16}{30}$

29. $\dfrac{2x}{5} = \dfrac{8}{10}$

30. $\dfrac{7x}{9} = \dfrac{12}{7}$

31. $\dfrac{7}{8}x = \dfrac{14}{24}$

32. $\dfrac{6}{5}x = \dfrac{18}{25}$

33. $\dfrac{12}{7}x = \dfrac{15}{14}$

34. $\dfrac{14}{27}x = \dfrac{21}{36}$

35. $\dfrac{18}{25}x = \dfrac{36}{75}$

36. $\dfrac{21}{32}x = \dfrac{35}{48}$

37. $\dfrac{12x}{25} = \dfrac{24}{75}$

38. $\dfrac{16x}{27} = \dfrac{24}{36}$

39. $\dfrac{18x}{35} = \dfrac{27}{21}$

40. $\dfrac{24x}{49} = \dfrac{36}{35}$

Section 6-2 Equations of the Form $x - a = b$ and $x + a = b$

Two laws of logic are given below; a, b, and c represent numbers.

> 1. If $a = b$, then $a + c = b + c$
> 2. If $a = b$, then $a - c = b - c$

In words, if the same number is added to (or subtracted from) two equal quantities, the resulting sums (or differences) are equal.

The first law can be used to solve the equation $x - 2 = 5$. The solution can be found by adding 2 to each side of the equation.

$$x - 2 = 5 \qquad\qquad Check: \quad 7 - 2 \stackrel{?}{=} 5$$
$$x - 2 + 2 = 5 + 2 \qquad\qquad\qquad\qquad 5 = 5$$
$$x = 7$$

When 2 is added to the left side of a balanced equation, 2 must also be added to the right side in order to maintain the balance.

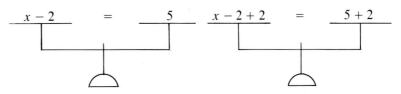

The equation below is solved using the second law.

$$x + 3 = 8 \qquad\qquad Check: \quad 5 + 3 \stackrel{?}{=} 8$$
$$x + 3 - 3 = 8 - 3 \qquad\qquad\qquad\qquad 8 = 8$$
$$x = 5$$

(Example continues on page 210.)

Complete the problems below. Completed problems appear on page 210.

Sample Set

	Equation:	Solution:
(a)	$x + 8 = 15$	$x + 8 - \Box = 15 - \Box$
		$x = \Box$
(b)	$x - 5 = 7$	$x - 5 + \Box = 7 + \Box$
		$x = \Box$

In order to maintain balance, 3 had to be subtracted from both sides of the equation.

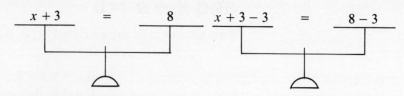

Laws of Logic

1. If $a = b$, then $a + c = b + c$
2. If $a = b$, then $a - c = b - c$
3. If $a = b$, then $a \times c = b \times c$
4. If $a = b$ and $c \neq 0$, then $\dfrac{a}{c} = \dfrac{b}{c}$

Completed Problems

Equation: Solution:

(a) $x + 8 = 15$ $x + 8 - \boxed{8} = 15 - \boxed{8}$

$x = \boxed{7}$

(b) $x - 5 = 7$ $x - 5 + \boxed{5} = 7 + \boxed{5}$

$x = \boxed{12}$

Name: _____

Class: _____

Exercises for Sec. 6-2 Solve each equation. Check your answers.

1. $x + 3 = 5$

2. $x - 2 = 4$

3. $x - 2 = 0$

4. $x + 5 = 5$

5. $x - 3 = 9$

6. $x + 5 = 8$

7. $x + 5 = 20$

8. $x - 6 = 20$

9. $x - 7 = 23$

10. $x + 4 = 17$

11. $x + 9 = 9$

12. $x - 9 = 0$

13. $x + 12 = 15$

14. $x - 15 = 21$

15. $x - 3.2 = 5.6$

16. $x + 2.8 = 4.5$

17. $x + 2.7 = 6.9$

18. $x - 3.5 = 9.2$

19. $x + \dfrac{3}{7} = \dfrac{5}{7}$

20. $x - \dfrac{2}{5} = \dfrac{4}{5}$

21. $x - \dfrac{2}{3} = \dfrac{2}{3}$

22. $x + \dfrac{2}{7} = \dfrac{6}{7}$

23. $x + \dfrac{1}{2} = \dfrac{3}{4}$

24. $x - \dfrac{2}{3} = \dfrac{5}{6}$

25. $x - \dfrac{3}{5} = \dfrac{4}{15}$

26. $x + \dfrac{3}{8} = \dfrac{3}{4}$

27. $2.5 + x = 3.6$
    (same as $x + 2.5 = 3.6$)

28. $3.2 + x = 8.5$
    (same as $x + 3.2 = 8.5$)

29. $\dfrac{5}{12} + x = \dfrac{11}{18}$

30. $\dfrac{2}{15} + x = \dfrac{3}{10}$

Name: _____

Class: _____

Unit 6 Test Solve each equation. Check your answers.

1. $2x = 6$

2. $3x = 7$

3. $15x = 20$

4. $18x = 54$

5. $\frac{1}{3}x = 5$

6. $\frac{2}{5}x = 8$

7. $\frac{1}{3}x = \frac{5}{6}$

8. $\frac{2}{5}x = \frac{3}{10}$

9. $\frac{2x}{5} = 3$

10. $\frac{5x}{7} = 10$

11. $\frac{3x}{8} = \frac{5}{11}$

12. $\frac{5x}{12} = \frac{35}{36}$

13. $x - 5 = 8$

14. $x + 3 = 17$

15. $x + 8.2 = 9.6$

16. $x - 2.7 = 4.5$

17. $x + \dfrac{5}{12} = \dfrac{3}{4}$

18. $x - \dfrac{5}{12} = \dfrac{3}{12}$

19. $x + \dfrac{1}{3} = \dfrac{3}{4}$

20. $x - \dfrac{1}{6} = \dfrac{2}{9}$

Section 7-1 Percent as a Number

The word **percent** means *hundredths*. The number to the left of the percent symbol (%) tells us how many hundredths we have.

$$1\% = \frac{1}{100}$$
$$19\% = \frac{19}{100}$$
$$100\% = \frac{100}{100} = 1$$
$$200\% = \frac{200}{100} = 2$$

Using the definition of percent, two useful conversion formulas can be obtained. See below.

$$N\% = \frac{N}{100} \quad \text{by definition}$$
$$= N \div 100$$
$$= N \times \frac{1}{100} \quad \text{used when converting to a common fraction}$$
$$= N \times .01 \quad \text{used when converting to a decimal}$$

Examples are given on pages 218 and 219.

Complete the problem below. Completed problems appear on page 219.

Sample Set

(a) $15\% = 15 \times .01 = \boxed{}$ (decimal) $= \boxed{}$ (fraction)

(b) $\frac{2}{9}\% = \frac{2}{9} \times \frac{1}{100} = \boxed{}$ (fraction) $= \boxed{}$ (decimal)

(c) $21.5\% = \boxed{} \times .01 = \boxed{}$ (decimal) $= \boxed{}$ (fraction)

(d) $5\frac{2}{3}\% = \boxed{} \times \frac{1}{100} = \boxed{}$ (fraction) $= \boxed{}$ (decimal)

(e) $.29 = .29 \times 100\% = \boxed{}\%$

(f) $\frac{2}{3} = \frac{2}{3} \times 100\% = \boxed{}\%$ (mixed) $= \boxed{}\%$ (decimal)

(g) $.782 = \boxed{} \times 100\% = \boxed{}\%$ (decimal) $= \boxed{}\%$ (mixed)

(h) $\frac{1}{375} = \boxed{} \times 100\% = \boxed{}\%$ (fraction) $= \boxed{}\%$ (decimal)

Examples

> Percent to common fractions:
>
> (a) $5\% = 5 \times \frac{1}{100} = \frac{5}{100} = \frac{1}{20}$
> (b) $\frac{2}{3}\% = \frac{2}{3} \times \frac{1}{100} = \frac{2}{300} = \frac{1}{150}$
> (c) $12\frac{5}{6}\% = 12\frac{5}{6} \times \frac{1}{100} = \frac{77}{6} \times \frac{1}{100} = \frac{77}{600}$
>
> Percent to decimal:
>
> (d) $5\% = 5 \times .01 = .05$
> (e) $23.2\% = 23.2 \times .01 = .232$
> (f) $549\% = 549 \times .01 = 5.49$
> (g) $.03\% = .03 \times .01 = .0003$

In the examples that follow, each percent is represented as a decimal and a common fraction.

Examples

> (a) $7.5\% = 7.5 \times .01 = .075$
>
> $7.5\% = 7.5 \times \frac{1}{100} = \frac{7.5}{100} = \frac{75}{1,000}$
>
> (b) $2\frac{3}{8}\% = 2\frac{3}{8} \times .01 = 2.375 \times .01 = .02375$
>
> $2\frac{3}{8}\% = 2\frac{3}{8} \times \frac{1}{100} = \frac{19}{8} \times \frac{1}{100} = \frac{19}{800}$
>
> (c) $15\frac{2}{3}\% = 15.666\cdots \times .01 = .15666\cdots$
>
> $15\frac{2}{3}\% = 15\frac{2}{3} \times \frac{1}{100} = \frac{47}{3} \times \frac{1}{100} = \frac{47}{300}$

Which representation is best? There is no definite answer. The form you choose may depend on which one is easiest to find, or on the problem you are working.

To convert a number to a percent, multiply the number by $\frac{100}{100}$ ($= 1$). Here are two examples:

$$.05 = \frac{.05}{1} \times \frac{100}{100}$$

$$= \frac{5}{100} = 5\% \qquad \text{recall that \% means hundredths}$$

$$\frac{1}{2} = \frac{1}{\cancel{2}} \times \frac{\cancel{100}^{50}}{100}$$

$$= \frac{50}{100} = 50\%$$

Sec. 7-1 Percent as a Number

Realizing that $\frac{100}{100} = 100$ percent $= 1$, the more convenient short form may be used.

Examples

(a) $.05 = .05 \times 1 = .05 \times 100\% = 5\%$
(b) $.003 = .003 \times 100\% = .3\%$
(c) $2.6 = 2.6 \times 100\% = 260\%$
(d) $\frac{1}{2} = \frac{1}{2} \times 100\% = (\frac{1}{2} \times 100)\% = 50\%$
(e) $\frac{3}{8} = \frac{3}{8} \times 100\% = (\frac{3}{8} \times 100)\% = 37\frac{1}{2}\%$
(f) $\frac{2}{3} = \frac{2}{3} \times 100\% = (\frac{2}{3} \times 100)\% = 66\frac{2}{3}\%$

Certain problems are simple enough so that conversions can be made mentally.

Examples

$.07 = 7$ hundredths $= 7\%$
$.326 = 32.6$ hundredths $= 32.6\%$
$9\% = 9$ hundredths $= .09$
$7.3\% = 7.3$ hundredths $= .073$

Completed Problems

(a) $15\% = 15 \times .01 = \boxed{.15}^{\text{decimal}} = \boxed{\frac{15}{100}}^{\text{fraction}}$

(b) $\frac{2}{9}\% = \frac{2}{9} \times \frac{1}{100} = \boxed{\frac{2}{900}}^{\text{fraction}} = \boxed{.00222}^{\text{decimal}}$

(c) $21.5\% = \boxed{21.5} \times .01 = \boxed{.215}^{\text{decimal}} = \boxed{\frac{215}{1000}}^{\text{fraction}}$

(d) $5\frac{2}{3}\% = \boxed{\frac{17}{3}} \times \frac{1}{100} = \boxed{\frac{17}{300}}^{\text{fraction}} = \boxed{.0567}^{\text{decimal}}$

(e) $.29 = .29 \times 100\% = \boxed{29}\%$

(f) $\frac{2}{3} = \frac{2}{3} \times 100\% = \boxed{66\frac{2}{3}}^{\text{mixed}}\% = \boxed{66.67}^{\text{decimal}}\%$

(g) $.782 = \boxed{.782} \times 100\% = \boxed{78.2}^{\text{decimal}}\% = \boxed{78\frac{1}{5}}^{\text{mixed}}\%$

(h) $\frac{1}{375} = \boxed{\frac{1}{375}} \times 100\% = \boxed{\frac{100}{375}}^{\text{fraction}}\% = \boxed{.2667}^{\text{decimal}}\%$

Name: _____

Class: _____

Exercises for Sec. 7-1   Represent each percent as a decimal.

1. 7% 2. 9%

3. 39% 4. 53%

5. 5.2% 6. 3.7%

7. 83.7% 8. 64.8%

9. .6% 10. .7%

11. 185% 12. 251%

13. 456.3% 14. 373.8%

15. $7\frac{1}{2}$% 16. $9\frac{1}{4}$%

17. $5\frac{3}{4}$% 18. $6\frac{2}{5}$%

19. $8\frac{2}{3}$%
(to the nearest 10,000th) 20. $5\frac{5}{6}$%
(to the nearest 10,000th)

Represent each percent as a common fraction.

21. $\tfrac{3}{4}\%$ 22. $\tfrac{3}{5}\%$

23. $2\tfrac{1}{3}\%$ 24. $3\tfrac{2}{9}\%$

25. $5\tfrac{2}{5}\%$ 26. $7\tfrac{3}{4}\%$

27. $8\tfrac{5}{8}\%$ 28. $9\tfrac{3}{8}\%$

29. $12\tfrac{3}{4}\%$ 30. $15\tfrac{3}{4}\%$

31. $125\tfrac{1}{2}\%$ 32. $250\tfrac{1}{3}\%$

33. 5% 34. 7%

35. 173% 36. 127%

37. 5.2% 38. 6.9%

39. .9% 40. .7%

Sec. 7-1 Percent as a Number

Represent each number as a percent in which the number in front of the % sign is a decimal. For example,

$$\tfrac{2}{3} = \tfrac{2}{3} \times 100\% = 66\tfrac{2}{3}\% = 66.7\% \quad \leftarrow \textbf{decimal to the nearest 10th}$$

41. .5   42. .7

43. .72   44. .39

45. .06   46. .05

47. .529   48. .738

49. .078   50. .092

51. .0075   52. .0082

53. $\tfrac{2}{5}$ 54. $\tfrac{3}{4}$

55. $\tfrac{3}{8}$ 56. $\tfrac{5}{8}$

57. $\tfrac{1}{3}$ 58. $\tfrac{1}{6}$

59. $\tfrac{2}{7}$ 60. $\tfrac{5}{9}$

Represent each number as a percent in which the number in front of the % sign is a mixed number or common fraction. For example,

$$.252 = .252 \times 100\% = 25.2\% = 25\tfrac{1}{5}\% \quad \leftarrow \textbf{mixed number}$$

61. $\tfrac{1}{2}$ 62. $\tfrac{1}{5}$

63. $\tfrac{3}{8}$ 64. $\tfrac{5}{8}$

65. $\tfrac{1}{3}$ 66. $\tfrac{1}{7}$

67. $\tfrac{1}{40}$ 68. $\tfrac{1}{50}$

69. $\tfrac{1}{200}$ 70. $\tfrac{1}{125}$

71. $\tfrac{1}{250}$ 72. $\tfrac{1}{400}$

73. .002 74. .003

75. .025 76. .076

77. .253 78. .397

79. .3547 80. .5869

Section 7-2 Basic Problems with Percents

There are three basic quantities in every problem with percents—the **base**, the **percent**, and the **percentage**—as indicated below.

$$\underset{\text{percentage}}{20} = \underset{\text{percent}}{25\%} \times \underset{\text{base}}{80}$$

There are three basic types of questions that can be asked. Each involves finding one of these quantities when the other two are known.

Type 1. **What** is 20 percent of 80? (percent and base known)

Type 2. 18 is **what** percent of 72? (percentage and base known)

Type 3. 30 is 15 percent of **what** number? (percentage and percent known)

Each of these questions can be translated into an equation. The word *is* is equivalent to *equal* and can be replaced with an = sign. The words *what* or *what number* represent a number to be found and can be represented with an *x*. The word *of* indicates multiplication. Since a multiplication sign might be confused with the *x* representing the unknown, parentheses should be used to indicate multiplication.

Examples with Questions of Type 1

(a) What is 20 percent of 80?

$x = (20\%)(80)$
$x = (20)(.01)(80)$ recall that $20\% = 20 \times .01 = (20)(.01)$
$x = (.20)(80)$
$x = 16$

Conclusion: 16 is 20 percent of 80.

(b) $5\frac{2}{3}$ percent of 75 is what number?

$$\left(5\frac{2}{3}\%\right)(75) = x$$

$$\left(\frac{17}{3}\%\right)(75) = x$$

$$\left(\frac{17}{\cancel{3}}\right)\left(\frac{1}{\cancel{100}}\right)(\cancel{75}) = x \qquad \text{Recall that } \frac{17}{3}\% = \frac{17}{3} \times \frac{1}{100}$$
$$= \left(\frac{17}{3}\right)\left(\frac{1}{100}\right)$$

$$\frac{17}{4} = x$$

$$4\frac{1}{4} = x$$

Conclusion: $5\frac{2}{3}$ percent of 75 is $4\frac{1}{4}$.

226 Percents

The question in part *b* above may not appear to be of the same form as the question in part *a*. It can be seen to be of the same form when replaced by the equivalent question:

"What is $5\frac{2}{3}$ percent of 75?"

Examples with Questions of Type 2

(a) 18 is what percent of 72?

$$18 = (x\%)(72)$$
$$18 = (x)(.01)(72)$$
$$18 = x(.72)$$
$$\frac{18}{.72} = \frac{x(\cancel{.72})}{\cancel{.72}} \quad \text{to get } x \text{ alone, divide each side by .72}$$
$$25 = x \quad\quad 18 \div .72 = 25$$

Conclusion: 18 is 25 percent of 72.

(b) What percent is 36 of 150?
Before writing the equation, **reword the question so that the words *what percent* appear in the center.**
The reworded question is:

in the center
36 is what percent of 150?

$$36 = (x\%)(150)$$
$$36 = x(.01)(150)$$
$$\frac{36}{1.5} = \frac{x(\cancel{1.5})}{\cancel{1.5}}$$
$$24 = x$$

Conclusion: 36 is 24 percent of 150.

Examples with Questions of Type 3

(a) 30 is 15 percent of what number?

$$30 = (15\%)(x)$$
$$30 = (15)(.01)x$$
$$\frac{30}{.15} = \frac{\cancel{.15}x}{\cancel{.15}}$$
$$200 = x$$

Conclusion: 30 is 15 percent of 200.

(b) If 12 is 180 percent of a number, what is the number?
Equivalent question: "12 is 180 percent of what number?"

$$12 = (180\%)(x)$$
$$12 = (180)(.01)x$$
$$\frac{12}{1.8} = \frac{\cancel{1.8}}{\cancel{1.8}} x$$
$$6\frac{2}{3} = x \qquad\qquad \frac{12}{1.8} = \frac{12}{1.8} \times \frac{10}{10} = \frac{\cancel{120}^{20}}{\cancel{18}_{3}} = 6\frac{2}{3}$$

Conclusion: 12 is 180 percent of $6\frac{2}{3}$.

Answer the following questions. Answers appear on page 228.

Sample Set

Translate each question into an equation.

(a) What is 75 percent of 84?

(b) 50 is what percent of 90?

(c) 72 is 120 percent of what number?

Which question is equivalent to the given question?

(d) What percent is 21 of 24?
  (1) 24 is what percent of 21?
  (2) 21 is what percent of 24?

(e) What number is 4 equal to 6 percent of?
  (1) 4 is 6 percent of what number?
  (2) What is 6 percent of 4?

(f) What number is 275 equal to 120 percent of?
  (1) What is 120 percent of 275?
  (2) 275 is 120 percent of what number?

(g) If 70 percent of a number is 52, what is the number?
  (1) 52 is 70 percent of what number?
  (2) What is 70 percent of 52?

Answers to Questions

(a) $x = (75\%)(84)$ (b) $50 = (x\%)(90)$ (c) $72 = (120\%)(x)$
(d) 2   (e) 1   (f) 2   (g) 1

Exercises for Sec. 7-2 Answer each question.

1. What is 5 percent of 90?

2. What is 6 percent of 70?

3. 15 is what percent of 20?

4. 14 is what percent of 50?

5. 45 is 5 percent of what number?

6. 32 is 4 percent of what number?

7. If 5 is 20 percent of a number, what is the number?

8. If 6 is 30 percent of a number, what is the number?

9. 6½ percent of 80 is what number?

10. 5¾ percent of 40 is what number?

11. 12 is 250 percent of what number?

12. 15 is 150 percent of what number?

13. What is ⅔ percent of 60?

14. What is ⅚ percent of 42?

15. 15 is what percent of 500?

16. 8 is what percent of 400?

17. What percent is 75 of 200?

18. What percent is 60 of 300?

19. If 5 percent of a number is 3, what is the number?

20. If 6 percent of a number is 12, what is the number?

21. What percent is 2 of 300?

22. What percent is 3 of 150?

23. 250 percent of 80 is what number?

24. 320 percent of 70 is what number?

25. What is 125 percent of 80?

26. What is 250 percent of 50?

27. What number is 15 equal to 25 percent of?

28. What number is 18 equal to 30 percent of?

29. If 150 percent of a number is 90, what is the number?

30. If 250 percent of a number is 75, what is the number?

31. If 6 is 120 percent of a number, what is the number?

32. If 30 is 150 percent of a number, what is the number?

33. 5 is $\frac{2}{3}$ percent of what number?

34. 8 is $\frac{2}{7}$ percent of what number?

35. 5 percent of 50 is what number?

36. 6 percent of 60 is what number?

37. What percent is 200 of 125?

38. What percent is 300 of 120?

39. What number is 30 equal to 150 percent of?

40. What number is 40 equal to 160 percent of?

Section 7-3 Application of Percent

There is no set way to work all word problems. In terms of the information given, the student must ask a question, in a familiar form, that is equivalent to the question being asked in the problem.

Sample Problem 7-1

A store owner decides to mark down all merchandise 25 percent. How much will he mark down a coat that is listed at $85?
Information: List price = $85
                Markdown = 25% of the list price
Question: What is 25 percent of $85?
$$x = (25\%)(\$85)$$
$$x = (25)(.01)(\$85)$$
$$x = (.25)(\$85)$$
$$x = \$21.25$$
Conclusion: The coat will be marked down $21.25.

Sample Problem 7-2

A sample of ore weighing 350 pounds contains 29 pounds of lead. What percent of the sample is lead?
Information: Sample weight = 350 lb
                Lead in sample = 29 lb
Question: 29 pounds is what percent of 350 pounds?
$$29 = (x\%)(350)$$
$$29 = x(.01)(350)$$
$$\frac{29}{3.5} = \frac{x(3.5)}{3.5} \qquad \frac{29}{3.5} = \frac{\cancel{290}^{58}}{\cancel{35}_{7}} = 8\frac{2}{7}$$
$$8\frac{2}{7} = x$$
Conclusion: 29 pounds is $8\frac{2}{7}$ percent of 350 pounds.

Sample Problem 7-3

A retailer buys a coat for $24 and sells it for $30. Margin is the difference obtained when the cost is subtracted from the selling price. What is the percent of margin? (Figure margin as a percent of cost.)
Information: Cost = $24
                Selling price = $30
                Margin = $30 − $24 = $6
Question: Margin is what percent of list, or $6 is what percent of $24?
$$6 = (x\%)(24)$$
$$6 = x(.01)(24)$$
$$\frac{6}{.24} = \frac{x(\cancel{.24})}{\cancel{.24}}$$
$$25 = x$$
Conclusion: $6 is 25 percent of $24, or margin is 25 percent of cost.

Percents

Sample Problem 7-4

A certain kind of ore contains $27\frac{1}{2}$ percent iron. How much ore is needed to obtain 572 pounds of iron?

Information: We want 572 lb of iron.
　　　　　　　Iron is $27\frac{1}{2}\%$ of the ore.

Question: 572 pounds is $27\frac{1}{2}$ percent of how many pounds?

$$572 = \left(27\frac{1}{2}\%\right)(x)$$

$$572 = (27.5)(.01)x$$

$$\frac{572}{.275} = \frac{.275x}{.275}$$

$$2{,}080 = x \qquad\qquad 572 \div .275 = 2{,}080$$

Conclusion: 572 pounds is $27\frac{1}{2}$ percent of 2,080 pounds, or 2,080 pounds of ore is required.

Sample Problem 7-5

A tradesman made $10.40 per hour in 1975 and made 6 percent more per hour in 1976. What was his hourly rate in 1976?

Information: 1975 hourly rate = $10.40
　　　　　　　He made 6% more than $10.40 in 1976.

You must conclude that in 1976 he made $10.40 + 6 percent of $10.40.

First question: What is 6 percent of $10.40?

$$x = (6\%)(\$10.40)$$
$$x = (6)(.01)(\$10.40)$$
$$x = (.06)(\$10.40)$$
$$x = \$.62 \quad\text{to the nearest cent}$$

Second question: What is $10.40 + 6 percent of $10.40?
Answer: $10.40 + $.62 = $11.02

Name: _____

Class: _____

Exercises for Sec. 7-3

1. An inspector tested 22 pounds of hamburger and found that $32\frac{1}{2}$ percent was cereal matter. How many pounds of cereal matter were in the sample? (Round your answer to the nearest 10th of a pound.)

2. A store owner had a gross income of $125,000. She spent $5,000 on advertising. What percent of her gross income was spent on advertising?

3. A smelting company got 327 pounds of pure metal from 1 ton (2,000 pounds) of ore. What percent of the ore was pure metal? (Round to the nearest 10th of 1 percent.)

4. In an election, 27,970 people voted. It was reported on the news that 30.4 percent of the eligible voters had voted. How many eligible voters were there? (Round to the nearest thousand.)

5. A woman had $24,000 to invest. She invested $10,000 in stocks and the rest in bonds.
   (a) What percentage did she invest in stocks?
   (b) What percentage did she invest in bonds?
   (Round to the nearest 10th of 1 percent.)

6. If a new car depreciated 42 percent the first year, how much was the first-year depreciation if the car cost $6,800?

7. One kilometer is approximately 1,094 yards. One mile is 1,760 yards. A kilometer is what percent of a mile? (Round to the nearest whole percent.)

8. A salesperson receives a commission of 8 percent of the selling price of each item sold. How much is the commission if an item sold for $49.95? (Round to the nearest cent.)

9. One thousand people were surveyed, and 429 were found to be Republicans. What percent of the people surveyed were Republicans? (Round to the nearest 10th of 1 percent.)

10. An inspector found that 5 percent of the refrigerators produced were defective. About how many refrigerators would she have to inspect in order to find 20 defective ones?

11. The cost of a certain product went up $8. If the original cost was $20, what percent did the cost increase? (Increase is figured as a percent of the original cost.)

12. An analysis of a 25-pound sample of brass showed that it contained 16.35 pounds of pure copper. What percent of the sample was pure copper? (Round to the nearest whole percent.)

13. An alloy was found to be 32 percent pure zinc. How much pure zinc was in 400 pounds of the alloy?

14. If 8 quarts of antifreeze are put into a 24-quart radiator and the remainder is filled with water, what percent of the solution will be antifreeze?

15. A manufacturer tested 1,000 bearings and found that 2.3 percent were defective. How many defective bearings did he find?

16. A consumer spent $25 for groceries, of which $12.75 was for meat. What percent of the purchase was for meat?

17. A dealer claims that 92.7 percent of the seedcorn he sells will germinate. Out of every 1,000 seeds planted, what number might be expected to grow?

18. A manufacturer found that 72 percent of the cost of production was for labor. If the cost of production was $42,500,000, how much of that cost was for labor?

19. Approximately 20 percent of the calories taken into the body are retained. About how many calories are retained when 1,500 calories are taken in?

20. A salesperson told a customer that the price of cameras was going up 20 percent next week. If a camera is presently priced at $250, how much will its price increase?

21. A company pays a dividend of $3\frac{2}{5}$ percent on each share of stock, where the percent is figured on the market value at the end of the fiscal year. If the value of the stock was $32.50 at the end of the fiscal year, what was the dividend? (Round to the nearest cent.)

22. The interest a bank paid each year was $5\frac{3}{4}$ percent of the amount deposited. How much interest would be earned if $3,200 were deposited?

23. A portable television set has a list price of $219.95. If the price is marked down 25 percent, how much will the television cost? (Round to the nearest cent.)

24. A company's 1975 annual report stated that their earnings increased 510 percent over the $1,900,000 earned in 1974. How much did they earn in 1975? (1975 earnings = 1974 earnings + 510 percent of 1974 earnings. Round to the nearest 100,000.)

25. A company's 1975 gross profit increased 30 percent over 1974. If the gross profit in 1974 was $56,000,000, what was the gross profit in 1975? (1975 gross profit = 1974 gross profit + 30 percent of 1974 gross profit. Round to the nearest million.)

26. At a clearance sale, a sofa with list price $430 was marked down to $300. What percent was the sofa marked down? (The amount of markdown is figured as a percentage of list price. Round to the nearest 10th of 1 percent.)

27. The population of a city was 42,000 persons according to the 1960 census, and 65,000 according to the 1970 census. What percent did the population increase? (The increase is figured as a percentage of the beginning population. Round to the nearest whole percent.)

28. A salesperson told a manufacturer that a punch press could be purchased that would increase output to 120 percent of the present output. If the present output is 72 items per day, how many items could be produced with the new press? (Round to the nearest whole number.)

29. In 1954, the average weight of a lineman on the Upstate College football team was 185 pounds. In 1975, the average weight was 245 pounds. What percent did the weight increase? (Increase is figured as a percentage of beginning weight. Round to the nearest 10th of 1 percent.)

30. A manufacturer found that manufacturing costs rose 7.5 percent in the last year. If her product cost $3 last year, how much more should she sell it for this year in order to break even?
(*Hint:* She needs 7.5 percent more than she got last year. Round to the nearest cent.)

31. A store manager is told that a certain item must be marked up $33\frac{1}{3}$ percent above cost. If the item costs $48, what should the item be sold for?

32. In 1974, a farmer's income was $32,000. In 1975, his income was $40,000. What percent did his income increase? (The increase is figured as a percent of beginning income.)

33. A company's annual report stated that 48 percent of their net sales were in chemicals and photoplates. If their net sales were $2,500,000, what dollar amount was due to chemicals and photoplates?

34. A company reported their tax rate to be 48 percent. This means their tax is 48 percent of their earnings. If their tax was $27,000,000, what were their earnings? (Round to the nearest million.)

35. The penalty for not paying income tax when it is due is ½ percent of the unpaid amount each month the tax is overdue. If a taxpayer owes $540 and pays 1 month late, what is the penalty?

36. If a metal beam expands .00065 percent of its length for each degree rise in temperature, how much will a 100-foot beam expand if the temperature rises:
    (a) 1 degree?
    (b) 3.5 degrees?
    (c) 70 degrees?
    (Round each answer to the nearest 1,000,000th.)

37. A chemical company bought 52 percent of the stock of a mining company. If the value of the stock they acquired was $20,800,000, what was the value of all the stock of the mining company?

38. A taxpayer found that her income tax should be $3,260 plus 28 percent of the excess of her taxable income over $16,000. Her taxable income was $18,500.
    (a) What was the excess over $16,000?
    (b) What was 28 percent of the excess?
    (c) What was her total tax?

39. In a certain town, the assessed value of a house is 25 percent of the true market value. The tax rate is $4.02 for each $100 of assessed value.
    (a) What is the assessed value of a house with a market value of $50,000?
    (b) How many hundreds of dollars was the assessed value? (Number of hundreds = assessed value ÷ 100.)
    (c) What was the tax?

40. A man borrowed $40,000 from a bank at a rate of 9 percent interest per year. He must pay back $360 each month until the loan is repaid.
    (a) How much interest does he owe at the end of the first month? (*Hint:* If he owes 9 percent of $40,000 for 1 year, he owes $\frac{1}{12}$ of 9 percent of $40,000 for 1 month.)
    (b) Subtract the first month's interest from the $360 payment to find how much principal will be subtracted from the debt.
    (c) How much does he owe after he makes the first payment?
    (d) Using the balance found in part c, repeat parts a, b, and c for the second month.

Name: _____

Class: _____

Unit 7 Test   Represent each percent as a decimal.

1. 27%
2. 8.4%
3. $5\frac{2}{5}\%$

Represent each percent as a common fraction.

4. $3\frac{1}{3}\%$
5. 2.3%

Represent each number as a percent in which the number in front of the % sign is a decimal.

6. .023
7. $\frac{5}{9}$ (nearest 10th of 1%)
8. $\frac{1}{250}$

Represent each number as a percent in which the number in front of the % sign is a fraction.

9. .243
10. $\frac{7}{8}$
11. $\frac{1}{500}$

Answer each of the following questions.

12. 12 is 30 percent of what number?

13. 20 is what percent of 160?

14. What is 36 percent of 90?

245

15. If 20 is 40 percent of a number, what is the number?

16. What percent is 15 of 40?

17. 80 percent of 120 is what number?

18. A 200-pound sample of ore is found to contain 12 pounds of pure lead. What percent of the sample is pure lead?

19. If a certain kind of ore is known to contain 6 percent pure copper, how much ore would be needed in order to obtain 300 pounds of pure copper?

20. If 6 percent of a certain ore is pure iron, how much iron could be obtained from 2,000 pounds of ore?

Section 8-1 Negative Numbers and Inequalities

What would be the outcome in each of the situations described below?

1. The temperature is 3 degrees and goes down 5 degrees.
2. You buy an item for $5 and sell it for $3.

The outcome in the first example would be a temperature of 2 degrees below zero and could be represented as -2 degrees. In the second example, the outcome would be a loss of $2 and could be represented as $-\$2$.

$$3° - 5° = -2°$$
$$\$3 - \$5 = -\$2$$
$$3 - 5 = -2$$

The number -2 is called a **negative** number and is read "minus 2," "negative 2," or "the negative of 2." The numbers without signs that you have worked with so far are positive numbers and can be represented with a $+$ sign. Positive and negative numbers are represented on the number line below.

$$-5 \quad -4 \quad -3 \quad -2 \quad -1 \quad 0 \quad +1 \quad +2 \quad +3 \quad +4 \quad +5$$

A number that appears without a sign will be understood to be positive.

$$1 = +1 \quad \text{and} \quad 3.2 = +3.2$$

In each of the examples above, the outcome could be viewed as representing a number less than zero.

$$-2° \text{ is } less \text{ than } 0°$$

There is less heat present when the temperature is -2 degrees than when the temperature is 0 degrees.

$$-\$2 \text{ is } less \text{ than } \$0$$

The person with $-\$2$ ($2 in the hole) has less (is worse off) than the person with $\$0$. Similarly, a person with $-\$5$ has less than a person with $-\$2$.

The symbols $<$ and $>$ are used to indicate the situations of less than and greater than, as illustrated below.

Less than

The Symbols	Are Read as
$3 < 8$	3 is less than 8
$0 < 3$	0 is less than 3
$-2 < 0$	-2 is less than 0
$-5 < -2$	-5 is less than -2

Arithmetic of Signed Numbers

Greater than

The Symbols	Are Read as
8 > 3	8 is greater than 3
3 > 0	3 is greater than 0
0 > −2	0 is greater than −2
−2 > −5	−2 is greater than −5

Note that the arrow points from the greater number to the lesser number. The statements above are called **inequalities**. The symbols < and > are called **inequality signs**.

As one moves to the right on the number line, the numbers become greater; and as one moves to the left, the numbers become less.

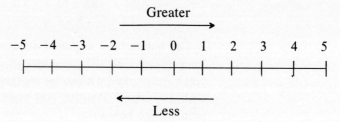

Sample Problem 8-1

Represent the less than and greater than relationship between the numbers −4 and 2 using the inequality signs, < and >.

Solution: $-4 < 2, \quad 2 > -4$

Sample Problem 8-2

Represent the order of the numbers $3\frac{1}{3}$, $-2\frac{1}{2}$, 0, 5, and −7 using < and >.

Solution:
$$-7 < -2\frac{1}{2} < 0 < 3\frac{1}{3} < 5$$
$$5 > 3\frac{1}{3} > 0 > -2\frac{1}{2} > -7$$

Place the correct inequality sign, < or >, in each listing of numbers. Completed problems appear on page 252.

Sample Set

(a) 7 ☐ 2 (b) 5 ☐ −2

(c) −3 ☐ 0 (d) −2 ☐ 5

(e) −5 ☐ 2 (f) −7 ☐ −3

(g) 2 ☐ 7 ☐ 9 (h) 5 ☐ 2 ☐ −3

(i) −800 ☐ −2 ☐ 1 ☐ 7,502

(j) −2 ☐ −3⅓ ☐ −5.1 ☐ −100

Completed Problems

(a) $7 \boxed{>} 2$ (b) $5 \boxed{>} -2$

(c) $-3 \boxed{<} 0$ (d) $-2 \boxed{<} 5$

(e) $-5 \boxed{<} 2$ (f) $-7 \boxed{<} -3$

(g) $2 \boxed{<} 7 \boxed{<} 9$ (h) $5 \boxed{>} 2 \boxed{>} -3$

(i) $-800 \boxed{<} -2 \boxed{<} 1 \boxed{<} 7{,}502$

(j) $-2 \boxed{>} -3\tfrac{1}{3} \boxed{>} -5.1 \boxed{>} -100$

Name: _____

Class: _____

Exercises for Sec. 8-1 Place the correct inequality sign, < or >, between each pair of numbers.

1. 5 3
2. 3 5
3. 0 2

4. 0 −2
5. −5 2
6. −5 −7

7. 3 −4
8. −7 −2
9. −6 −9

10. 7 −10
11. −8 −2
12. −100 1

13. $\frac{1}{3}$ $\frac{1}{2}$
14. $-\frac{1}{3}$ $\frac{1}{4}$
15. $-\frac{3}{4}$ $-\frac{1}{2}$

16. $\frac{1}{5}$ $-\frac{2}{3}$
17. 3.2 3.25
18. −7.1 −3.2

19. 4.6 −5.9
20. $-2\frac{1}{2}$ $\frac{1}{3}$

Arrange the numbers in order so that the inequality sign < can be used. Place the sign < between the numbers.

21. 7, 1, 5, 12
22. −2, −3, −1, −7

23. −2, 5, 12, 0, −6
24. 8, −4, 2, −15, −3

25. −5, −7, 3, −4, 1, −1, 7
26. −8, 7, −6, −2, −7, 6, 4, 3, −3

27. −.9, −.82, .3, .31, −.05, −.21, .19

28. $\frac{1}{3}, \frac{1}{10}, \frac{3}{5}, \frac{1}{100}, \frac{8}{9}$
29. $-\frac{1}{6}, \frac{1}{10}, -\frac{1}{5}, -\frac{3}{4}, \frac{1}{4}, \frac{5}{7}$

30. $.6, \frac{1}{4}, -\frac{2}{5}, -.1, -.7, .59, 0, -.001$

254 Arithmetic of Signed Numbers

Arrange the numbers in order so that the inequality sign > can be used. Place the sign > between the numbers.

31. 4, 7, 1, 5

32. −7, −2, −5, −1

33. −3, 2, −5, 0, 7

34. −5, 2, 7, −6, 1

35. 7, −4, −3, −5, 2, 1, −1

36. .2, −.16, −.2, .7, .16, −.1, −.23

37. −2, 3, −5, 0, −4, −1, 4, 1, 9

38. $-\frac{1}{4}, -\frac{2}{5}, -\frac{9}{10}, -\frac{1}{20}, -\frac{3}{50}$

39. $-\frac{1}{2}, \frac{5}{6}, \frac{1}{3}, -\frac{5}{8}, -\frac{8}{9}, \frac{2}{15}$

40. $\frac{1}{3}, -\frac{2}{5}, .1, -.8, -\frac{1}{10}, .01, 0, -.01$

Section 8-2 Addition and Subtraction of Two Signed Numbers

Subtraction of Positives and Addition of Negatives

Subtracting a positive number is equivalent to moving that many units to the left on the number line.

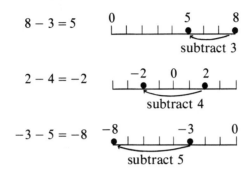

$8 - 3 = 5$

$2 - 4 = -2$

$-3 - 5 = -8$

Adding a negative number is also equivalent to moving to the left on the number line.

$8 + (-3) = 5$

$2 + (-4) = -2$

$-3 + (-5) = -8$

(*Note:* The parentheses are used only to group the sign as part of the number being added.)

The above examples illustrate that **subtracting a positive number gives the same result as adding the negative of that number.**

Subtract a positive	*Add a negative*
$8 - 3 = 5$	$8 + (-3) = 5$
$2 - 4 = -2$	$2 + (-4) = -2$
$-3 - 5 = -8$	$-3 + (-5) = -8$

Subtracting a positive 3 may be viewed as spending $3 for an item with money out of your pocket, that is, taking away $3 that you had. Adding a minus 3 may be viewed as charging a $3 item, that is, adding a debt (−$3). In either case, the result would be the same.

Addition of Positives and Subtraction of Negatives

Adding a positive number is equivalent to moving that many units to the right on the number line.

$3 + 5 = 8$ (add 5, from 0 to 8)

$-2 + 3 = 1$ (add 3, from −2 to 1)

$-8 + 4 = -4$ (add 4, from −8 to −4)

Subtracting a negative number is also equivalent to moving to the right on the number line.

$3 - (-5) = 8$ (subtract −5, from 0 to 8)

$-2 - (-3) = 1$ (subtract −3, from −2 to 1)

$-8 - (-4) = -4$ (subtract −4, from −8 to −4)

The examples indicate that **subtracting a negative is equivalent to adding a positive.** This will be discussed further.

Subtract a negative:	*Add a positive:*
$3 - (-5) = 8$	$3 + 5 = 8$
$-2 - (-3) = 1$	$-2 + 3 = 1$
$-8 - (-4) = -4$	$-8 + 4 = -4$

A Key Point   **Every subtraction problem is equivalent to an addition problem.**

Subtraction	*Equivalent addition*
$A - B$	$A + (-B)$
$A - (-B)$	$A + B$

Justification in Terms of Inverse Operations

It was stated earlier that subtraction is the inverse operation of addition. Addition is also the inverse operation of subtraction. That is,

If, $c - b = a$ is true
Then $a + b = c$ must be true

As an example,

If $8 - 3 = 5$ is true
Then $5 + 3 = 8$ must be true

Sec. 8-2 Addition and Subtraction of Two Signed Numbers

This approach can be used to find the answer to the subtraction problem $4 - (-3)$. Let x represent the answer. Then,

If $\quad 4 - (-3) = x \quad$ is true
Then $\quad x + (-3) = 4 \quad$ must be true

Or, using the commutative law,

$$-3 + x = 4 \quad \text{must be true}$$

Starting at -3, what x must be added to -3 in order to get 4? It is easy to see that x must be 7. We therefore have

$$4 - (-3) = 7$$

Complete the following problems. Completed problems appear on page 258.

Sample Set

Place the correct sign here ↓   Answers ↓

(a) $3 + (-5) = 3 \;\square\; 5 = \square$

(b) $-4 - 6 = -4 \;\square\; (-6) = \square$

(c) $7 - 4 = 7 \;\square\; (-4) = \square$

(d) $-3 + (-4) = -3 \;\square\; 4 = \square$

(e) $-5 + 8 = -5 \;\square\; (-8) = \square$

(f) $2 - (-7) = 2 \;\square\; 7 = \square$

(g) $-5 - (-3) = -5 \;\square\; 3 = \square$

(h) $7 + 3 = 7 \;\square\; (-3) = \square$

A Special Sum

Observe the following sum:

$$2 + (-2) = 0$$

The number -2 is called the **additive inverse** of 2 since it is the number which, when added to 2, gives the sum 0. The notion of inverse (or opposite) comes from the fact that an addition of -2 "undoes" an addition of 2.

$$5 + 2 + (-2) = 5 + 0$$
$$= 5$$

We started with 5 and added 2. Adding -2 got us back to 5.

Completed Problems

(a) $3 + (-5) = 3 \boxed{-} 5 = \boxed{-2}$

(b) $-4 - 6 = -4 \boxed{+} (-6) = \boxed{-10}$

(c) $7 - 4 = 7 \boxed{+} (-4) = \boxed{3}$

(d) $-3 + (-4) = -3 \boxed{-} 4 = \boxed{-7}$

(e) $-5 + 8 = -5 \boxed{-} (-8) = \boxed{3}$

(f) $2 - (-7) = 2 \boxed{+} 7 = \boxed{9}$

(g) $-5 - (-3) = -5 \boxed{+} 3 = \boxed{-2}$

(h) $7 + 3 = 7 \boxed{-} (-3) = \boxed{10}$

Exercises for Sec. 8-2 Find each sum or difference.

1. $-2 - (-4)$
2. $-2 + (-4)$

3. $-2 + 4$
4. $-2 - 4$

5. $6 + (-3)$
6. $7 - (-4)$

7. $-5 + 8$
8. $-5 + 2$

9. $-6 - 9$
10. $-6 + 9$

11. $3 - 8$
12. $-7 - (-10)$

13. $-8 - (-8)$
14. $9 - (-12)$

15. $13 + (-18)$
16. $-18 + 13$

17. $-4 - (-3)$
18. $-5 + (-5)$

19. $10 + (-10)$
20. $2 - (-9)$

21. $3 - (-2)$
22. $-15 - (-5)$

23. $-14 - (-15)$
24. $7 + (-15)$

25. $-15 - (-7)$ 26. $-12 + (-10)$

27. $-11 + 5$ 28. $12 - 19$

29. $27 - (-20)$ 30. $-27 + (-20)$

Section 8-3 Addition and Subtraction Problems Involving Several Numbers

In the problem $3 - 4 + 8 - 5$, positive numbers are being added and subtracted. Each sign is a sign of operation. The problem is worked below.

$$3 - 4 + 8 - 5 = 11 - 9 \qquad Note: \quad 3 + 8 = 11$$
$$= 2 \qquad\qquad\qquad\quad -4 - 5 = -9$$

The combining of numbers preceded by like signs depends on using the commutative and associative laws of addition. In order to see that these laws apply, we must write the problem as an addition problem. This is shown below.

$3 - 4 + 8 - 5 = 3 + (-4) + 8 + (-5)$ written as an addition problem

$\qquad\qquad\qquad = (3 + 8) + ((-4) + (-5))$ ← commutative and associative laws used here to rearrange and group numbers

$\qquad\qquad\qquad = 11 + (-9)$

$\qquad\qquad\qquad = 11 - 9$

$\qquad\qquad\qquad = 2$

The process of combining numbers preceded by like signs may be viewed as combining gains and losses. For example, suppose we had gains of $3 and $8, and losses of $4 and $5. Then the total of our gains would be $11, and the total of our losses would be $9. Subtracting would show the outcome to be a $2 overall gain. (Sample problems are given on page 262.)

Complete the following problems. Completed problems appear on page 262.

Sample Set

Place the correct signs in each box.

(a) $\quad -3 + (-4) - (-6) + 7 = -3 \,\square\, 4 \,\square\, 6 + 7$

(b) $\quad -8 + 5 - 7 + 9 = -8 \,\square\, (-5) \,\square\, (-7) + 9$

(c) $\quad 5 - 6 - 8 + 7 = 5 \,\square\, 7 \,\square\, 6 \,\square\, 8 \qquad$ rearranged

$\qquad\qquad\qquad\quad = \,\square \qquad\qquad\qquad\qquad$ answer

(d) $\quad -6 - (-9) + (-4) + 5 = -6 \,\square\, 9 \,\square\, 4 + 5$

$\qquad\qquad\qquad\qquad\quad = \,\square\, 9 \,\square\, 5 \,\square\, 6 \,\square\, 4 \quad$ rearranged

$\qquad\qquad\qquad\qquad\quad = \,\square \qquad\qquad\qquad\qquad$ answer

262 Arithmetic of Signed Numbers

Sample Problem 8-3

Express $-5 + 6 - (-3) - 2$ as an addition problem.

Solution:

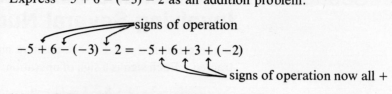

$$-5 + 6 - (-3) - 2 = -5 + 6 + 3 + (-2)$$

(*Note:* Subtracting -3 is equivalent to adding 3. Subtracting 2 is the same as adding -2.

Sample Problem 8-4

Compute $8 - 7 + (-4) - (-6)$.

Solution:

Method 1 (as a problem in which positive numbers are added and subtracted):

$$8 - 7 + (-4) - (-6)$$
$$= 8 - 7 - 4 + 6$$
$$= 8 + 6 - 7 - 4$$
$$= 14 - 11$$
$$= 3$$

Method 2 (entirely as an addition problem):

$$8 - 7 + (-4) - (-6)$$
$$= 8 + (-7) + (-4) + 6$$
$$= (8 + 6) + (-7) + (-4)$$
$$= 14 + (-11)$$
$$= 3$$

Note the similarity in the steps. Which method you use depends on which you find easiest.

Completed Problems

(a) $\quad -3 + (-4) - (-6) + 7 = -3 \boxed{-} 4 \boxed{+} 6 + 7$

(b) $\quad -8 + 5 - 7 + 9 = -8 \boxed{-} (-5) \boxed{+} (-7) + 9$

(c) $\quad 5 - 6 - 8 + 7 = 5 \boxed{+} 7 \boxed{-} 6 \boxed{-} 8$

$\qquad\qquad\qquad\qquad = \boxed{-2} \quad$ answer

(d) $\quad -6 - (-9) + (-4) + 5 = -6 \boxed{+} 9 \boxed{-} 4 + 5$

$\qquad\qquad\qquad\qquad = \boxed{+} 9 \boxed{+} 5 \boxed{-} 6 \boxed{-} 4$

$\qquad\qquad\qquad\qquad = \boxed{4} \quad$ answer

Name: _____

Class: _____

Exercises for Sec. 8-3 Write each problem as an addition problem.

1. $-2 + 5 - 6$
2. $5 - 7 - 8$
3. $6 - 5 + 7 - 8$
4. $-7 - 8 + 6 - 2$
5. $4 - 3 + (-4)$
6. $-6 + (-4) - 5$
7. $-5 + (-2) - 3 - (-1)$
8. $9 - (-2) - 5 + 7$
9. $-3 - (-5) - 8 - 4 + (-6)$
10. $-5 + 6 - (-5) - (-2) - 7$

Compute each answer.

11. $-2 + 5 - 7$

12. $5 - 8 + 2$

13. $5 - 6 - 7 + 2$

14. $-6 + 8 - 7 + 3$

15. $-7 - 8 - 9 - 10$

16. $-5 - 6 - 7 - 12$

17. $4 - 7 - 8 + 9 - 2$

18. $-5 + 3 + 9 - 7 + 2$

19. $-5 - 8 - 7 + 6 - 5 + 2 + 8$

20. $5 - 8 - 7 - 9 + 3 - 7 + 10$

21. $-7 + 8 - 3 - 8 + 7 + 4 - 9$

22. $2 - 7 + 8 + 5 + 6 - 8 - 10 + 3$

23. $5 - (-2) + (-3)$

24. $-6 - (-2) + 7$

25. $-7 - (-2) + (-3) - (-2)$

26. $6 - (-3) - 8 + (-4)$

27. $-7 + (-3) + 8 - (-4) + (-5)$

28. $6 - 9 - (-7) - (-6) - 3$

29. $-14 - (-3) + (-6) - (-7) - 6 + (-4)$

30. $-5 - (-7) - 15 + (-6) + 6 - (-5)$

31. $3 - (-3) - 3 + (-3)$

32. $5 + (-2) - 5 - (-2)$

33. $-7 + (-3) - (-6) + 8 - (-7)$

34. $8 - (-7) + (-6) - 4 + 8 - 10$

35. $-15 + (-6) - (-3) - 4 + (-5)$

36. $7 - (-5) + (-7) + (-3) - 6 - (-3)$

37. $4 - 6 + (-8) + (-5) - (-2) + (-7)$

38. $-7 - (-6) + 5 - (-8) + (-3) - (-5)$

39. $5 - (-3) + (-7) - (-8) + (-4) + (-9)$

40. $-7 + (-8) - (-2) - (-7) + (-6) - (-3)$

Section 8-4 Multiplication and Division of Signed Numbers

The product of two numbers with opposite signs is negative.

Examples

$$(2)(-3) = -6$$
$$(-7)(5) = -35$$

The first problem may be viewed as doubling a $3 debt. The result would be a $6 debt, thus negative. This result can also be found using the idea that multiplication is a shortcut for addition.

$$2(-3) = -3 + (-3) = -6$$

If you agree that a positive times a negative should be negative, the product in the second example is justified by the commutative law.

$$(-7)(5) = (5)(-7) = -35$$

The product of two negative numbers is positive.

Examples

$$(-2)(-3) = 6$$
$$(-3)(-3) = 9$$

This rule will appear reasonable when we observe the pattern of the multiples of -3 shown below.

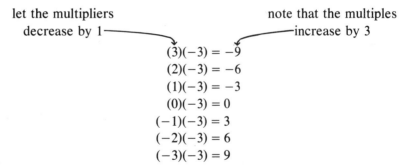

let the multipliers decrease by 1   note that the multiples increase by 3

$$(3)(-3) = -9$$
$$(2)(-3) = -6$$
$$(1)(-3) = -3$$
$$(0)(-3) = 0$$
$$(-1)(-3) = 3$$
$$(-2)(-3) = 6$$
$$(-3)(-3) = 9$$

When numbers with opposite signs are divided, the quotient is negative.

Examples

$$\frac{-6}{2} = -3$$
$$\frac{8}{-4} = -2$$

The first example may be viewed as the division of a $6 debt into two equal parts. Each part would be a $3 debt.

268 Arithmetic of Signed Numbers

It will be shown below that both of the above quotients are negative. The fact that division is the inverse operation of multiplication will be used.

Inverse Statement $\qquad$ If $A \times B = C \quad$ then $\dfrac{C}{B} = A$

In particular, the statement below is true.

If $\quad (-3)(2) = -6 \quad$ is true

then $\quad \dfrac{-6}{2} = -3 \quad$ must be true

Since $(-3)(2) = -6$ is true, then $\dfrac{-6}{2} = -3$ is also true.

Also, if $\quad (-2)(-4) = 8 \quad$ is true

then $\quad \dfrac{8}{-4} = -2 \quad$ must be true

Since $(-2)(-4) = 8$ is true, then $\dfrac{8}{-4} = -2$ is also true.

The quotient obtained when two negative numbers are divided is positive.

Examples
$$\dfrac{-6}{-2} = 3$$
$$\dfrac{-12}{-6} = 2$$

Again, the fact that division is the inverse operation of multiplication can be used.

Rule $\qquad$ If $A \times B = C \quad$ then $\dfrac{C}{B} = A$

In particular,

if $\quad (2)(-6) = -12 \quad$ is true

then $\quad \dfrac{-12}{-6} = 2 \quad$ must be true

Since $(2)(-6) = -12$ is true, $\dfrac{-12}{-6} = 2$ is also true.

Sec. 8-4 Multiplication and Division of Signed Numbers

Products with More than Two Numbers

Do you see a relationship between the number of negative factors and the sign of the answer in the problems below?

$$(-2)(-3) = +6$$

$$\underbrace{(-2)(-3)}_{+6}(-5) = (+6)(-5) = -30$$

$$\underbrace{(-2)(-3)(-5)}_{-30}(-2) = (-30)(-2) = +60$$

$$\underbrace{(-2)(-3)(-5)(-2)}_{+60}(-4) = (+60)(-4) = -240$$

Note: When the number of negative factors being multiplied is even, the answer is positive. When the number of negative factors is odd, the product is negative.

This rule for counting minus signs can be extended to problems in which both multiplication and division are performed.

Examples

(a) $\dfrac{-4}{-2} = +2$ two − signs answer: +

(b) $\dfrac{(-4)(-3)}{-2} = \dfrac{+12}{-2} = -6$ three − signs answer: −

(c) $\dfrac{-24}{(-2)(-3)} = \dfrac{-24}{+6} = -4$

(d) $\dfrac{(-4)(-3)(-2)}{-6} = \dfrac{-24}{-6} = +4$ four − signs answer: +

(e) $\dfrac{(-12)(-3)}{(-2)(-9)} = \dfrac{+36}{+18} = +2$

(f) $\dfrac{(-12)(-3)(-4)}{(-9)(-2)} = \dfrac{-144}{+18} = -8$ five − signs answer: −

(g) $\dfrac{(-3)(-4)(-6)(-2)}{-12} = \dfrac{+144}{-12} = -12$

Arithmetic of Signed Numbers

Cancellation may be used in problems of the type just shown. **The safest thing to do is** first determine the sign of the numerator and denominator, and **rewrite the problem** as shown in the example below.

Sample Problem 8-5

Compute $\dfrac{(-4)(+9)(-2)}{(+6)(-3)}$.

Solution: Determine the sign of the numerator and denominator, and rewrite the problem; then cancel:

$$\dfrac{\overbrace{(-4)(+9)(-2)}^{+}}{\underbrace{(+6)(-3)}_{-}} \qquad \dfrac{+(\cancel{4})(9)(\cancel{2})}{-(\cancel{6})(\cancel{3})} = -4$$

Note that counting the minus signs can be used to check the sign of the final answer.

Warning The rule for counting minus signs should not be used in problems in which addition or subtraction is performed.

Two minus signs appear in each of the problems below. However, the signs of the answers are different.

$$\dfrac{\overbrace{(-2)(+3)}^{-6} + 8}{(-6)(+2)} = \dfrac{-6+8}{-12} = \dfrac{+2}{-12} = -\dfrac{1}{6}$$

$$\dfrac{\overbrace{(+2)(+3)}^{+6} + 8}{(-6)(-2)} = \dfrac{+6+8}{+12} = \dfrac{+14}{+12} = +\dfrac{7}{6}$$

Problems of this type will be covered in the next unit.

Sec. 8-4 Multiplication and Division of Signed Numbers

In the problems below, place the correct sign in each box. Completed problems appear on page 272.

Sample Set

(a) $(-2)(-4) = \boxed{}8$ (b) $(-7)(+8) = \boxed{}56$

(c) $\dfrac{-12}{4} = \boxed{}3$ (d) $\dfrac{24}{-3} = \boxed{}8$

(e) $\dfrac{-15}{-5} = \boxed{}3$ (f) $(-3)(-4)(6) = \boxed{}72$

(g) $(-5)(-2)(-3) = \boxed{}30$ (h) $(-4)(-2)(5)(-2) = \boxed{}80$

(i) $(-3)(-4)(-5)(-4) = \boxed{}240$

(j) $\dfrac{(-2)(8)(-3)}{(-6)(-4)} = \dfrac{\boxed{}\,(\cancel{2})(\cancel{8})(\cancel{3})}{\boxed{}\,(\cancel{6})(\cancel{4})} = \boxed{}2$

(k) $\dfrac{(-10)(-18)(-12)}{(-8)(-27)(15)} = \dfrac{\boxed{}\,(\cancel{10})(\cancel{18})(\cancel{12})}{\boxed{}\,(\cancel{8})(\cancel{27})(\cancel{15})} = \boxed{}\dfrac{2}{3}$

Justification of the Equality
$A \times (B - C) = A \times B - A \times C$

Recall that the equality $A \times (B - C) = A \times B - A \times C$ was used to justify the subtraction rule for fractions (see page 105). The numbers A, B, and C may be considered positive.

$$\begin{aligned} A \times (B - C) &= A \times (B + (-C)) \\ &= (A \times B) + (A \times (-C)) \\ &= A \times B + (-(A \times C)) \\ &= A \times B - A \times C \end{aligned}$$

like $5 - 2 = 5 + (-2)$
Distributive law
like $5 \times (-2) = -(5 \times 2)$
like $7 + (-3) = 7 - 3$

Completed Problems

(a) $(-2)(-4) = \boxed{+}\,8$

(b) $(-7)(+8) = \boxed{-}\,56$

(c) $\dfrac{-12}{4} = \boxed{-}\,3$

(d) $\dfrac{24}{-3} = \boxed{-}\,8$

(e) $\dfrac{-15}{-5} = \boxed{+}\,3$

(f) $(-3)(-4)(6) = \boxed{+}\,72$

(g) $(-5)(-2)(-3) = \boxed{-}\,30$

(h) $(-4)(-2)(5)(-2) = \boxed{-}\,80$

(i) $(-3)(-4)(-5)(-4) = \boxed{+}\,240$

(j) $\dfrac{(-2)(8)(-3)}{(-6)(-4)} = \dfrac{\boxed{+}\,(\cancel{2})(\cancel{8})^{\,1}_{\,\cancel{2}}(\cancel{3})^{1}}{\boxed{+}\,(\cancel{6})_{\cancel{2}}^{\,1}(\cancel{4})_{1}} = \boxed{+}\,2$

(k) $\dfrac{(-10)(-18)(-12)}{(-8)(-27)(15)} = \dfrac{\boxed{-}\,(\cancel{10})^{2}(\cancel{18})^{\,1}_{\,\cancel{2}}(\cancel{12})^{\,1}_{\,\cancel{4}}}{\boxed{+}\,(\cancel{8})^{\,1}_{\,\cancel{2}}(\cancel{27})_{3}(\cancel{15})_{\cancel{3}}^{\,1}} = \boxed{-}\,\dfrac{2}{3}$

Name: _____

Class: _____

Exercises for Sec. 8-4 Find the products.

1. $(-5)(-3)$ 2. $(-2)(4)$ 3. $(3)(-4)$

4. $(-7)(8)$ 5. $(-3)(4)$ 6. $(-6)(-5)$

7. $(8)(-6)$ 8. $(7)(-9)$ 9. $(-8)(-9)$

10. $(-10)(-8)$

11. $(-3)(-4)(-2)$

12. $(-4)(-5)(-3)$

13. $(5)(-2)(3)$

14. $(5)(3)(-4)$

15. $(-6)(-2)(3)$

16. $(-5)(2)(-3)$

17. $(2)(-5)(-3)(2)$

18. $(-3)(4)(-2)(3)$

19. $(-2)(-6)(3)(-2)$

20. $(-3)(4)(-2)(-3)$

21. $(-2)(-2)(5)(-2)(-3)$

22. $(-5)(-4)(2)(-3)(3)$

23. $(-2)(3)(-2)(3)(-2)$

24. $(5)(-2)(-3)(-2)(-2)$

Find each quotient.

25. $\dfrac{-12}{6}$ 26. $\dfrac{-9}{3}$ 27. $\dfrac{-24}{-8}$

28. $\dfrac{-25}{-5}$ 29. $\dfrac{18}{-3}$ 30. $\dfrac{24}{-3}$

31. $\dfrac{-12}{8}$ 32. $\dfrac{-15}{10}$ 33. $\dfrac{-30}{-24}$

34. $\dfrac{-48}{-36}$ 35. $\dfrac{54}{-36}$ 36. $\dfrac{56}{-24}$

Compute the answers.

37. $\dfrac{(8)(-3)}{(-4)(9)}$ 38. $\dfrac{(-6)(4)}{(-8)(9)}$

39. $\dfrac{(-6)(-8)}{(-12)(5)}$ 40. $\dfrac{(9)(-8)}{(-12)(-4)}$

41. $\dfrac{(-8)(-9)}{(12)(-6)(-3)}$ 42. $\dfrac{(8)(-9)}{(-12)(6)(-3)}$

43. $\dfrac{(-12)(-9)(-4)}{(-8)(-6)}$ 44. $\dfrac{(-16)(-6)(-12)}{(-8)(-9)}$

45. $\dfrac{(-20)(8)(-9)}{(-15)(-16)(-6)}$ 46. $\dfrac{(-24)(-9)(-6)}{(27)(-12)(-8)}$

47. $\dfrac{(-8)(-6)(9)(-4)}{(-27)(24)}$ 48. $\dfrac{(-12)(9)(-6)(8)}{(-36)(-16)}$

49. $\dfrac{(-5)(-8)(6)(-9)}{(-12)(-16)(10)}$ 50. $\dfrac{(-8)(-12)(9)(6)}{(-32)(27)(3)}$

Name: _____

Class: _____

Unit 8 Test Place the correct inequality sign, < or >, between the numbers.

1. −2   5 2. −2   −7

Arrange the numbers in the order for which the inequality sign < may be used. Place < between the numbers.

3. −2, 7, 4, −6, 0, −3, 1

Find the sum or difference.

4. $5 + (-3)$ 5. $8 - 9$ 6. $-2 + (-5)$

7. $-7 + 3$ 8. $9 - (-4)$ 9. $-6 - 3$

10. $-8 - (-3)$

Find the products.

11. $(-9)(-6)$ 12. $(-8)(5)$

13. $(-2)(3)(-4)(-1)$

Find the quotients.

14. $\dfrac{-30}{6}$ 15. $\dfrac{28}{-7}$ 16. $\dfrac{-24}{-16}$

Compute the answers.

17. $\dfrac{(-2)(-9)(-4)}{(8)(-6)}$

18. $\dfrac{(4)(-6)(-12)}{(9)(-8)(5)}$

19. $-2 + 6 - 9 - 4 + 7$

20. $-5 - (-6) + 8 + (-7) - 3 + 4$

Section 9-1 Problems Involving Multiplication, Division, Addition, and Subtraction

Examples of **terms** are given below.

$$2 \qquad \frac{2}{5} \qquad 3(2) \qquad 3\left(\frac{8}{2}\right) \qquad \frac{2(10)}{4}$$

The problem below has three terms. The statement of the problem is not a term because the operations of addition and subtraction are involved.

$$3(2) + 3\left(\frac{8}{2}\right) - \frac{2(10)}{4} \qquad \text{not a term}$$

In order to solve the problem, the number value of each term must first be computed. The operations of addition and subtraction are then performed. This order follows the general rule given below.

> **Multiplication and division precede addition and subtraction.**

Here is the solution to the above problem:

$$3(2) + 3\left(\frac{8}{2}\right) - \frac{2(10)}{4} = 6 + 3(4) - \frac{20}{4}$$
$$= 6 + 12 - 5$$
$$= 13$$

Examples

(a) $\dfrac{-12}{3}(2) - 4\left(\dfrac{3}{6}\right) + \dfrac{18}{2}(3) = -4(2) - 4\left(\dfrac{1}{2}\right) + 9(3)$
$\qquad\qquad\qquad\qquad\qquad\qquad = -8 - 2 + 27$
$\qquad\qquad\qquad\qquad\qquad\qquad = 17$

(b) $\dfrac{\frac{24}{3}}{2}(5) - 8\left(\dfrac{3}{6}\right)(2) = \dfrac{8}{2}(5) - 8\left(\dfrac{1}{2}\right)(2)$
$\qquad\qquad\qquad\qquad\qquad = 4(5) - 8$
$\qquad\qquad\qquad\qquad\qquad = 20 - 8$
$\qquad\qquad\qquad\qquad\qquad = 12$

Order of Operations

Complete the problems below.

Sample Set

(a) $\dfrac{15}{3} - (8)(3) + \dfrac{12}{4} = \square - \square + \square = \square$

(b) $\dfrac{-24}{6}(4) + \dfrac{(9)(2)}{6} - \dfrac{\frac{18}{3}}{2} + 4\left(\dfrac{3}{2}\right) = \square + \square - \dfrac{\square}{2} + \square$
$ = \square + \square - \square + \square$
$ = \square$

(c) $\dfrac{(8)(3)}{6}(2) - \dfrac{\frac{12}{3}(2)}{4} - \dfrac{18}{6}\left(\dfrac{4}{3}\right) = \square - \dfrac{\square}{4} - \square$
$ = \square - \square - \square = \square$

Completed Problems

(a) $\dfrac{15}{3} - (8)(3) + \dfrac{12}{4} = \boxed{5} - \boxed{24} + \boxed{3} = \boxed{-16}$

(b) $\dfrac{-\overset{4}{\cancel{24}}}{\cancel{6}}(4) + \dfrac{\overset{3}{\cancel{(9)(2)}}}{\underset{\cancel{3}}{\cancel{6}}} - \dfrac{\overset{6}{\cancel{\frac{18}{3}}}}{2} + 4\left(\dfrac{3}{2}\right) = \boxed{-16} + \boxed{3} - \dfrac{\boxed{6}}{2} + \boxed{6}$
$ = \boxed{-16} + \boxed{3} - \boxed{3} + \boxed{6}$
$ = \boxed{-10}$

(c) $\dfrac{\overset{4}{\cancel{(8)(3)}}}{\underset{2}{\cancel{6}}}(2) - \dfrac{\overset{4}{\cancel{\frac{12}{3}}}(2)}{4} - \dfrac{\cancel{18}}{\cancel{6}}\left(\dfrac{4}{3}\right) = \boxed{8} - \dfrac{\boxed{8}}{4} - \boxed{4}$
$ = \boxed{8} - \boxed{2} - \boxed{4} = \boxed{2}$

Exercises for Sec. 9-1 Compute each answer.

1. $\dfrac{12}{6} + 5(2)$

2. $\dfrac{18}{3} - 7(2)$

3. $-4(3) - \dfrac{12}{2}$

4. $5(4) - \dfrac{18}{3}$

5. $4(3) - 5(4)$

6. $-3(2) + \dfrac{12}{3}$

7. $\dfrac{12}{3} + 6(2) - \dfrac{15}{3}$

8. $\dfrac{18}{3} + \dfrac{12}{2} - (5)(2)$

9. $\dfrac{-24}{2} + \dfrac{12}{4} - 6(2)$

10. $6(3) - 5(4) + \dfrac{18}{2}$

11. $8(3) - 7(4) + \dfrac{15}{3}$

12. $-6(2) - \dfrac{15}{3} + \dfrac{20}{5}$

13. $\dfrac{16}{8}(3) + 5\left(\dfrac{6}{10}\right)$ 　　14. $\dfrac{20}{5}(2) - \dfrac{(6)(4)}{3}$

15. $\dfrac{4(6)}{8} - 6\left(\dfrac{2}{4}\right)$ 　　16. $\dfrac{15}{5}(2) - \dfrac{8(3)}{2}$

17. $\dfrac{\frac{-24}{3}}{2} + \dfrac{15}{3}(4)$ 　　18. $-8\left(\dfrac{2}{4}\right) - \dfrac{18}{2}\left(\dfrac{1}{3}\right)$

19. $\dfrac{-8(6)}{4} - \dfrac{15}{5}(3)$ 　　20. $\dfrac{18}{2}(2) - 5\left(\dfrac{4}{10}\right)$

21. $\dfrac{12}{3}(2) - 5(3)(2) + 8\left(\dfrac{3}{6}\right)$ 　　22. $\dfrac{15}{5}(2) - \dfrac{8(4)}{16} - 5\left(\dfrac{4}{10}\right)$

23. $\dfrac{9(4)}{6} + 6\left(\dfrac{3}{9}\right) - \dfrac{12}{3}(5)$ 　　24. $\dfrac{16}{8}(3) - \dfrac{18(5)}{9} + 3(2)(2)$

25. $\dfrac{-20(3)}{5} - 8\left(\dfrac{5}{4}\right) - \dfrac{6(2)}{4}$ 26. $\dfrac{-8(5)}{10} + \dfrac{12}{4}(3) - 15\left(\dfrac{2}{6}\right)$

27. $7\left(\dfrac{4}{2}\right) - \dfrac{9(2)}{6} + \dfrac{6(3)}{9}$ 28. $\dfrac{-18}{9}(2) - \dfrac{2(8)}{4} - \dfrac{30}{6}(2)$

29. $5(2)(3) - \dfrac{16}{8}(3) - \dfrac{15(2)}{3} + 3(2)(2)$

30. $\dfrac{15}{5}(2) + \dfrac{9(8)}{12} - \dfrac{18}{6}(5) - 9\left(\dfrac{2}{3}\right)$

31. $\dfrac{12}{4}(3) + \dfrac{15(2)}{5} - 8\left(\dfrac{3}{6}\right) + 5(2)(2)$

32. $6\left(\dfrac{4}{8}\right) - \dfrac{4(8)}{10} - \dfrac{16}{8}(3) + 4\left(\dfrac{6}{12}\right)$

33. $\dfrac{21}{7}(2) - \dfrac{4(5)}{2} - 3\left(\dfrac{6}{9}\right) - \dfrac{36}{6}(2)$

34. $\dfrac{24(5)}{8} - \dfrac{15}{5}(4) - 6\left(\dfrac{5}{10}\right) - 9\left(\dfrac{2}{3}\right)$

35. $\dfrac{-15}{3}(2) - \dfrac{7(4)}{14} + 8\left(\dfrac{5}{10}\right) + \dfrac{18}{9}(2)$

36. $\dfrac{-20(3)}{5} + 6\left(\dfrac{5}{10}\right) - \dfrac{18}{3}(5) + 8\left(\dfrac{4}{2}\right)$

37. $\dfrac{\tfrac{36}{6}(2)}{3} - \dfrac{8(3)}{6}(2) + \dfrac{\tfrac{16}{4}}{6}(3)$

38. $\dfrac{\tfrac{24}{8}}{9}(6) - 9\left(\dfrac{2}{6}\right)(3) - \dfrac{\tfrac{15}{5}(4)}{6}$

39. $\dfrac{-8(3)}{12}(3) + \dfrac{\frac{15}{3}(4)}{10} - 9\left(\dfrac{4}{6}\right)(2)$

40. $\dfrac{\frac{-18}{9}(6)}{3} - \dfrac{9(5)}{15}(2) + 6(4)\left(\dfrac{3}{12}\right)$

Section 9-2 Parentheses and Brackets Used to Group Operations

Recall that operations inside parentheses are to be performed first. Within parentheses, multiplication and division precede addition and subtraction.

Examples

(a) $5(3 - 2(6))$
$= 5(3 - 12)$ multiplication before subtraction
$= 5(-9)$
$= -45$

(b) $3 + 2(5 + 1)$
$= 3 + 2(6)$
$= 3 + 12$ multiplication before addition
$= 15$

(c) $(3 + (2)(3))\left(\dfrac{\cancel{6}^{2}}{\cancel{3}}(2)\right)$ parentheses used for grouping and to indicate multiplication
$= (3 + 6)(4)$
$= (9)(4)$
$= 36$

(d) $\left(\dfrac{\cancel{12}^{4}}{\cancel{3}}(2) + 7\right) \div \left(4 + \dfrac{3(\cancel{6})^{3}}{\cancel{2}}\right)$
$= (8 + 7) \div (4 + 9)$
$= 15 \div 13$
$= \dfrac{15}{13} = 1\dfrac{2}{13}$

Brackets [ ] serve the same purpose as parentheses. When parentheses are placed inside brackets, do the operations inside the parentheses first. Then do the remaining operations inside the brackets next.

Examples

(e) $8[5 + 2(3 - 7)]$
$= 8[5 + 2(-4)]$
$= 8[5 + (-8)]$ multiplication before addition
$= 8[5 - 8]$
$= 8[-3]$
$= -24$

(f) $\quad 5 + 2\left[6 - 2\left(\overset{2}{\cancel{\dfrac{\cancel{12}}{\cancel{6}}}}(3) - 10\right)\right]$ $\quad\overbrace{}^{\text{start here}}$

$= 5 + 2[6 - 2(6 - 10)]$
$= 5 + 2[6 - 2(-4)]$
$= 5 + 2[6 - (-8)]$ multiplication before subtraction
$= 5 + 2[6 + 8]$
$= 5 + 2[14]$
$= 5 + 28$ multiplication before addition
$= 33$

Complete the problems below.

Sample Set

(a) $6 - 2\left(5 + \dfrac{6}{3}\right)$

$= 6 - 2(5 + \boxed{})$

$= 6 - 2()$

$= 6 - \boxed{} = \boxed{}$

(b) $(5 - 4(2))\left(\dfrac{8}{4} - 5\right)$

$= (5 - \boxed{})(\boxed{} - 5)$

$= ()() = \boxed{}$

(c) $\dfrac{18}{3}\left(\overset{2}{\dfrac{\cancel{8}}{\cancel{4}}}(3) - 10\right)$

$= \boxed{}(\boxed{} - 10)$

$= \boxed{}() = \boxed{}$

(d) $6\left[5 - 2\left(3 - \dfrac{8}{2}\right)\right]$

$= 6[5 - 2(3 - \boxed{})]$

$= 6[5 - 2()]$

$= 6[5 - \boxed{}]$

$= 6\boxed{} = \boxed{}$

Completed Problems

(a) $6 - 2\left(5 + \dfrac{\cancel{6}^{2}}{\cancel{3}}\right)$

$= 6 - 2(5 + \boxed{2})$

$= 6 - 2(\boxed{7})$

$= 6 - \boxed{14} = \boxed{-8}$

(b) $(5 - 4(2))\left(\dfrac{8}{4} - 5\right)$

$= (5 - \boxed{8})(\boxed{2} - 5)$

$= (\boxed{-3})(\boxed{-3}) = \boxed{9}$

(c) $\dfrac{18}{3}\left(\dfrac{\cancel{8}^{2}}{\cancel{4}}(3) - 10\right)$

$= \boxed{6}(\boxed{6} - 10)$

$= \boxed{6}(\boxed{-4}) = \boxed{-24}$

(d) $6\left[5 - 2\left(3 - \dfrac{8}{2}\right)\right]$

$= 6[5 - 2(3 - \boxed{4})]$

$= 6[5 - 2(\boxed{-1})]$

$= 6[5 - \boxed{-2}]$

$= 6\boxed{7} = \boxed{42}$

Name: _____

Class: _____

Exercises for Sec. 9-2 Compute each answer.

1. $18(5 - 3)$
2. $8(5 + 2)$

3. $\dfrac{15}{5}(5 - 3)$
4. $5(4)(3 - 5)$

5. $\dfrac{14}{2}(2 + 3)$
6. $5(3 + 2(4))$

7. $7\left(5 - \dfrac{8}{4}\right)$
8. $7 - \left(2 + \dfrac{6}{3}\right)$

9. $6 - (6(2) - 7)$
10. $9 - 2(3(2) + 1)$

11. $8 - 3\left(\dfrac{6}{3} + 1\right)$
12. $\dfrac{18}{9}(3) - \left(10 - \dfrac{12}{2}\right)$

13. $2\left(\dfrac{15}{3} - 2\right) - \left(\dfrac{12}{6} + 3\right)$

14. $3(5(2) - 8) - (3(5) - 12)$

15. $\left(\dfrac{4}{2}(3) + 2\right)(5(3) - 10)$

16. $\left(5 - \dfrac{4(6)}{3}\right)\left(\dfrac{15}{3} - 8\right)$

17. $5 - [8 - (3 + 2)]$

18. $7 - \left(\dfrac{6}{2} + 5\right)$

19. $6 - [(3)(2) - (5 + 2)]$

20. $7 - \left[\dfrac{8}{4} - (5 + 3)\right]$

21. $7[5 + 2(3 + 1)]$

22. $6[5 - 3(2 + 1)]$

23. $8[(5)(2) - 2(3 + 4)]$

24. $9\left[\dfrac{6}{3} + 2(4 - 2)\right]$

25. $8 - 2[3 + 2(6 - 7)]$

26. $31 - 3[5 - 2(2 - 5)]$

27. $5 + 2[6(2) - 3(5(2) - 12)]$

28. $4 - 5\left[\dfrac{8}{2} + 2\left(\dfrac{6}{3} + 1\right)\right]$

29. $5[8 - 2(5 - 3)] - [7 - 3(6 - 3)]$

30. $6[9 - 3(2 + 2)] - [6(3 + 1) - 20]$

Section 9-3 Division Line

A division (fraction) line groups operations above and below the line. For example, the following problems are equivalent.

$$\frac{3 + 2(5)}{5 - \frac{6}{2}}$$

is the same as

$$(3 + 2(5)) \div \left(5 - \frac{6}{2}\right)$$

Complete the operations in the numerator and denominator before you divide.

Examples

(a) $\quad \dfrac{10 + 2}{3(2)} = \dfrac{12}{6} = 2$

Warning — **Do not cancel when operations other than multiplication appear in the numerator or denominator.**

(b) $\quad \dfrac{3 + 2(5)}{5 - \frac{6}{2}} = \dfrac{3 + 10}{5 - 3} = \dfrac{13}{2} = 6\frac{1}{2}$

(c) $\quad \dfrac{5 - (\overset{6}{\overbrace{3(2)}} - 8)}{\frac{8}{4} - \frac{15}{3}} = \dfrac{5 - (-2)}{2 - 5} = \dfrac{7}{-3} = -2\dfrac{1}{3}$

(d) $\quad 3 + \dfrac{\overset{3}{\cancel{18}}\,(2)}{3\underbrace{(2 + 1)}_{3}} = 3 + \dfrac{6}{9} = 3 + \dfrac{2}{3} = 3\dfrac{2}{3}$

(e) $\quad \dfrac{6\left(\overset{4}{\overbrace{\frac{8}{12}}}\right)}{4 + \underbrace{3(2)}_{6}} - \dfrac{4 - \frac{\overset{2}{\cancel{6}}}{3}}{\underbrace{\frac{15}{3}(2)}_{10}} = \dfrac{4}{10} - \dfrac{2}{10} = \dfrac{2}{10} = \dfrac{1}{5}$

Complete the problems below.

Sample Set

(a) $\dfrac{12-4}{2(4)} = \dfrac{\Box}{\Box} = \Box$

(b) $\dfrac{15-3(2)}{3+\dfrac{6}{2}} = \dfrac{15-\Box}{3+\Box} = \dfrac{\Box}{\Box} = \Box$

(c) $\dfrac{\overset{6}{\cancel{18}}}{\cancel{3}}(2) - 4 \over 4(6) - 6(2)} = \dfrac{\Box - 4}{\Box - \Box} = \dfrac{\Box}{\Box} = \dfrac{\Box}{\Box}$

(d) $\dfrac{\dfrac{\overset{6}{\cancel{36}}}{\cancel{6}}(2)}{\underset{\widetilde{18}}{6(3)-10}} - \dfrac{16 - \overset{9}{\widetilde{3(3)}}}{\dfrac{20}{5}(2)} = \dfrac{\Box}{\Box} - \dfrac{\Box}{\Box} = \dfrac{\Box}{\Box}$

Completed Problems

(a) $\dfrac{12-4}{2(4)} = \dfrac{\boxed{8}}{\boxed{8}} = \boxed{1}$

(b) $\dfrac{15-3(2)}{3+\frac{6}{2}} = \dfrac{15-\boxed{6}}{3+\boxed{3}} = \dfrac{\boxed{9}}{\boxed{6}} = \boxed{1\frac{1}{2}}$

(c) $\dfrac{\overset{6}{\cancel{18}}}{\cancel{3}}(2)-4 \over 4(6)-6(2)} = \dfrac{\boxed{12}-4}{\boxed{24}-\boxed{12}} = \dfrac{\boxed{8}}{\boxed{12}} = \dfrac{\boxed{2}}{\boxed{3}}$

(d) $\dfrac{\frac{\overset{6}{\cancel{36}}}{\cancel{6}}(2)}{\underset{18}{6(3)-10}} - \dfrac{16-\overset{9}{\cancel{3(3)}}}{\frac{20}{5}(2)} = \dfrac{\boxed{12}}{\boxed{8}} - \dfrac{\boxed{7}}{\boxed{8}} = \dfrac{\boxed{5}}{\boxed{8}}$

Name: _____

Class: _____

Exercises for Sec. 9-3 Compute each answer. Reduce fractional answers to lowest terms.

1. $\dfrac{8(3)}{2-10}$

2. $\dfrac{2-5}{3(2)}$

3. $\dfrac{5-7}{8-12}$

4. $\dfrac{8-5}{2-8}$

5. $\dfrac{5(2)-7}{\dfrac{16}{4}}$

6. $\dfrac{\dfrac{6}{3}+2}{3(2)}$

7. $\dfrac{\dfrac{27}{9}+7}{4-9}$

8. $\dfrac{\dfrac{12}{4}-5}{2-4}$

9. $\dfrac{\dfrac{18}{9} + 3}{2(4) - 10}$

10. $\dfrac{9(3) - 7}{12 - 3(2)}$

11. $\dfrac{5 - 3(4)}{\dfrac{6}{3} + 4}$

12. $\dfrac{7 + 3(3)}{\dfrac{8}{4} - 6}$

13. $\dfrac{4(2) - \dfrac{12}{4}}{\dfrac{18}{6} + 5(3)}$

14. $\dfrac{\dfrac{8}{2} + 3(2)}{5(3) - \dfrac{6}{2}}$

15. $\dfrac{\dfrac{28}{7} - 6(2)}{15 - \dfrac{12}{4}(2)}$

16. $\dfrac{\frac{6}{2} - 3\left(\frac{8}{6}\right)}{7\left(\frac{4}{2}\right) - 5(3)}$

17. $\dfrac{5 - (2(3) - 2)}{6 + 2(3 + 2)}$

18. $\dfrac{7 + 2(5 - 2)}{8 - 2\left(\frac{6}{3}\right)}$

19. $\dfrac{15 - 2(3 + 4)}{5(2)(3) - 6}$

20. $\dfrac{12 + 4(3 - 5)}{\frac{10}{5}(3) - 2}$

21. $\dfrac{7(3 - 2(3))}{5 - (10 - 4(3))}$

22. $\dfrac{(8 - 2(3))\left(\frac{6}{2}(3)\right)}{(3 + 7(3))\left(5 - \frac{6}{3}\right)}$

23. $\dfrac{\frac{3(4)}{6}(19 - 3(3))}{(5(4) - 8)(8(4) - 12)}$

24. $\dfrac{\left(7 - \dfrac{18}{2}\right)(2(3))}{\left(\dfrac{18}{6} + 3\right)\left(\dfrac{10}{5} - 4\right)}$

25. $5 + \dfrac{\dfrac{18}{6} + 3(2)}{3(4) + 3}$

26. $7 - \dfrac{\dfrac{16}{4} - 5(2)}{\dfrac{4(3)}{6} + 4}$

27. $\dfrac{2(7)}{2 + 4} - \dfrac{8 - 3}{3(2)}$

28. $\dfrac{5 - 3}{\dfrac{18}{3}} + \dfrac{\dfrac{16}{4} - 3}{\dfrac{12}{3}}$

29. $\dfrac{\dfrac{12}{3} + 2}{18 - 3(5)} - \dfrac{\dfrac{3}{4}(12)}{2(4) - 5}$

30. $\dfrac{5 - 2(3)}{\dfrac{15}{3} - 3} - \dfrac{5 + 3(4)}{8 - 3(2)}$

Unit 9 Test   Compute each answer. Reduce fractional answers to lowest terms.

1. $\dfrac{18}{3} - 5(2)$

2. $6\left(\dfrac{3}{9}\right) + \dfrac{3(4)}{6}$

3. $8(3) - 9(2) - \dfrac{16}{4}$

4. $-6\left(\dfrac{6}{9}\right) - \dfrac{3(4)}{2} + 3(2)(2)$

5. $\dfrac{12}{3}(6) - 2(5) - \dfrac{\frac{12}{4}(6)}{3}$

6. $\dfrac{8}{4}(2)(3) - \dfrac{4(12)}{3(6)}$

7. $-3\left(\dfrac{12}{6}\right)(4) - \dfrac{6\left(\dfrac{12}{3}\right)}{2(8)}$

8. $\dfrac{12}{3(2)}$

9. $5(4 + 2(3))$

10. $18 - 2(3 + 2)$

11. $(5 + 4)(3(5) - 8)$

12. $3[12 - 6(8 - 5)]$

13. $5 - [3 - 2(5)]$

14. $3(8 - 2(3)) - 4\left(3 + \dfrac{8}{2}\right)$

15. $\dfrac{8(2)}{5 - 3}$

16. $\dfrac{5(2) - 4}{6 - 3(3)}$

17. $\dfrac{12 - 2(3 + 1)}{\dfrac{8}{2} + 2}$

18. $\dfrac{\dfrac{16}{4}(4 + 2)}{3(5) - 7(4 + 5)}$

19. $\dfrac{8-3}{3(2)} - \dfrac{4(3)-9}{\dfrac{12}{2}}$

20. $2 + \dfrac{3(2)-4}{4+3(2)}$

Section 10-1 Basic Formulas

A **formula** is a rule stated as an equality using letters to represent numbers. An example is the formula for the area of a rectangle. The formula and the rule stated in words are given below.

$$A = l \times w$$
$$\text{Area} = \text{length} \times \text{width}$$

Here A represents the area, l the length, and w the width of the rectangle. If $l = 8$ and $w = 6$, then

$$A = 8 \times 6 = 48$$

When numbers replace letters in a formula, they are said to be **substituted** for the letters. The numbers that are substituted for letters will be referred to as **data**.

Letters placed next to each other indicate that the numbers they represent are to be multiplied. The area formula could thus have been written

$$A = lw$$

This is the same as

$$A = l \times w$$

Examples

(a) Compute the value of V in the formula $V = lwh$, given the data $l = 4$, $w = 3$, and $h = 5$.

$$V = lwh$$
$$V = (4)(3)(5) = 60$$

(b) Compute K in the formula $K = ab + c$, given $a = 2$, $b = 3$, and $c = 5$.

$$K = ab + c$$
$$K = (2)(3) + 5$$
$$= 6 + 5 \quad \textbf{multiplication}$$
$$= 11 \quad \textbf{before addition}$$

(c) Compute the value of J in the formula $J = a(b + c)$, given $a = 3$, $b = 4$, and $c = 3$.

$$J = a(b + c)$$
$$= 3(4 + 3)$$
$$= 3(7)$$
$$= 21$$

(d) Compute the value of L in the formula $L = (a + b)/ac$, given $a = 2$, $b = 10$, and $c = 3$.

$$L = \frac{a + b}{ac}$$

$$= \frac{2 + 10}{(2)(3)} = \frac{12}{6} = 2$$

(e) Compute the value of K in the formula $K = 3(r + s)/(rs + 2)$, given $r = 3$ and $s = 7$.

$$K = \frac{3(r + s)}{rs + 2}$$

$$= \frac{3(3 + 7)}{(3)(7) + 2} = \frac{3(10)}{21 + 2} = \frac{30}{23} = 1\frac{7}{23}$$

When the same number is repeated as a factor in a product, the product may be written in **exponential form**.

$$(3)(3) = 3^2 \quad \text{3 squared}$$
$$(3)(3)(3) = 3^3 \quad \text{3 cubed}$$
$$(3)(3)(3)(3) = 3^4 \quad \text{3 to the fourth power}$$
$$(3)(3)(3)(3)(3) = 3^5 \quad \text{3 to the fifth power}$$

base / exponent

Each exponential form above is a **power**; thus, 3^2 is a power. The number to which the power is taken is called the **base**. The **exponent** counts the number of times the base is used as a factor in the product. Below, several powers are computed using different bases.

$$2^2 = (2)(2) = 4 \qquad 5^2 = (5)(5) = 25$$
$$2^3 = (2)(2)(2) = 8 \qquad 5^3 = (5)(5)(5) = 125$$
$$2^4 = (2)(2)(2)(2) = 16 \qquad 5^4 = (5)(5)(5)(5) = 625$$

Examples

(a) Compute the value of R in the formula $R = ab^2$, given $a = 2$ and $b = 3$.

$$R = ab^2$$
$$= (2)(3)^2$$
$$= (2)(9) \qquad (3)^2 = (3)(3) = 9$$
$$= 18$$

Note that the exponent 2 applies only to $b = 3$. The number 3 was squared first. That product was then multiplied by 2.

(b) Compute the value of L in the formula $L = (ab)^2$, given $a = 2$ and $b = 3$.

$$L = (ab)^2$$
$$= ((2)(3))^2$$
$$= (6)^2$$
$$= 36$$

Since a and b are inside parentheses, the numbers were multiplied first; that product was then squared: $6^2 = (6)(6) = 36$.

(c) Compute the value of S in the formula $S = r^3(r+s)^2$, given $r = 3$ and $s = 2$.

$$S = r^3(r+s)^2$$
$$= 3^3(3+2)^2$$
$$= 3^3(5)^2$$
$$= 27(25) \qquad 3^3 = (3)(3)(3) = 27$$
$$= 675 \qquad 5^2 = (5)(5) = 25$$

(d) Compute the value of K in the formula $K = r^3/(s^2 + r)$, given $r = 5$ and $s = 4$.

$$K = \frac{r^3}{s^2 + r}$$
$$= \frac{5^3}{4^2 + 5}$$
$$= \frac{125}{16 + 5}$$
$$= \frac{125}{21} = 5\frac{20}{21}$$

Note that taking a power of a number is a multiplication operation and is, therefore, done before addition or subtraction.

Complete the problems below. Completed answers appear on page 315.

Sample Set

(a) $J = a(b - c) \qquad a = 3, b = 5, c = 3$

$= \square (\square - \square)$

$= \square (\square)$

$= \square$

(b) $K = ab - c^2 \qquad a = 2, b = 4, c = 5$

$= (\ )(\ ) - (\ )^2$

$= \square - \square$

$= \square$

(c) $L = a(b-a)^3 \quad a=3, b=5$

$= \Box(\Box - \Box)^3$

$= \Box(\Box)^3$

$= \Box(\Box)$

$= \Box$

(d) $J = \dfrac{a+b}{a-b} \quad a=3, b=5$

$= \dfrac{\Box + \Box}{\Box - \Box} = \dfrac{\Box}{\Box}$

$= \Box$

Completed Problems

(a) $J = a(b - c)$ $a = 3, b = 5, c = 3$

$= \boxed{3}(\boxed{5} - \boxed{3})$

$= \boxed{3}(\boxed{2})$

$= \boxed{6}$

(b) $K = ab - c^2$ $a = 2, b = 4, c = 5$

$= (\boxed{2})(\boxed{4}) - (\boxed{5})^2$

$= \boxed{8} - \boxed{25}$

$= \boxed{-17}$

(c) $L = a(b - a)^3$ $a = 3, b = 5$

$= \boxed{3}(\boxed{5} - \boxed{3})^3$

$= \boxed{3}(\boxed{2})^3$

$= \boxed{3}(\boxed{8})$

$= \boxed{24}$

(d) $J = \dfrac{a + b}{a - b}$ $a = 3, b = 5$

$= \dfrac{\boxed{3} + \boxed{5}}{\boxed{3} - \boxed{5}} = \dfrac{\boxed{8}}{\boxed{-2}}$

$= \boxed{-4}$

Exercises for Sec. 10-1 Find the value of K in each formula.

1. $K = abc \quad a = 2, b = 3, c = 4$

2. $K = rst \quad r = 3, s = 5, t = 2$

3. $K = a + b + c \quad a = 3, b = 7, c = 4$

4. $K = r + s + t \quad r = 2, s = 5, t = 6$

5. $K = a + bc \quad a = 2, b = 5, c = 4$

6. $K = rs + t \quad r = 3, s = 3, t = 7$

7. $K = a(b + c) \quad a = 3, b = 2, c = 5$

8. $K = r(s - t)$ $r = 3, s = 5, t = 2$

9. $K = 3n + 2m - 3$ $n = 4, m = 3$

10. $K = 2n + 3m + 5$ $n = 5, m = 2$

11. $K = ab + cb$ $a = 3, b = 2, c = 5$

12. $K = rs - st$ $r = 2, s = 5, t = 3$

13. $K = 5abc + 2$ $a = 3, b = 2, c = 2$

14. $K = 3rs + 2t$ $r = 3, s = 2, t = 5$

15. $K = (a - b)^2 \quad a = 5, b = 2$

16. $K = (r + s)^2 \quad r = 2, s = 3$

17. $K = (a + b)^3 \quad a = 3, b = 2$

18. $K = (r - s)^3 \quad r = 5, s = 2$

19. $K = a(b - a)^3 \quad a = 3, b = 5$

20. $K = r(r + s)^3 \quad r = 2, s = 3$

21. $K = \dfrac{m + n}{n} \quad m = 5, n = 3$

22. $K = \dfrac{x + y}{y}$ $x = 4, y = 2$

23. $K = \dfrac{mn + 3}{m + n}$ $m = 2, n = 3$

24. $K = \dfrac{xy + 3}{x - y}$ $x = 5, y = 3$

25. $K = mn^2$ $m = 2, n = 3$

26. $K = xy^3$ $x = 2, y = 3$

27. $K = (mn)^3$ $m = 2, n = 3$

28. $K = (xy)^2$ $x = 3, y = 4$

29. $K = m^2 n^3 \qquad m = 3, n = 2$

30. $K = x^2 y^3 \qquad x = 2, y = 3$

31. $K = \dfrac{r - s}{r^2} \qquad r = 5, s = 2$

32. $K = \dfrac{a + b}{a^2} \qquad a = 4, b = 3$

33. $K = \dfrac{r^2 + t}{r - s} \qquad r = 2, s = 4, t = 3$

34. $K = \dfrac{a^2 - b}{a + c} \qquad a = 5, b = 3, c = 3$

35. $K = \dfrac{3(r+s)}{r^2 - t}$ $r = 3, s = 2, t = 4$

36. $K = \dfrac{2(a-b)}{a^3 - c}$ $a = 3, b = 2, c = 7$

37. $K = \dfrac{r(r-s)}{s(r+s)}$ $r = 5, s = 1$

38. $K = \dfrac{a(b+a)}{b(a-b)}$ $a = 2, b = 3$

39. $K = \dfrac{r(s+t)^2}{s^3}$ $r = 3, s = 2, t = 4$

40. $K = \dfrac{a(b-c)^3}{c^2}$ $a = 3, b = 5, c = 2$

Section 10-2 Problems Involving More than One Formula

Quite often a value obtained from one formula is used to compute a value in a second formula. This is demonstrated in the problem below.

Sample Problem 10-1 Compute the value of W using the formulas $W = ab$, $a = m + n$, and $b = mn$, given the data $m = 3$ and $n = 2$.

Strategy: In order to find W in the formula $W = ab$, the values of a and b must be known. These values can be found by substituting $m = 3$ and $n = 2$ into the formulas $a = m + n$ and $b = mn$.

Solution:
$$m = 3 \qquad n = 2$$
$$a = \overset{3}{m} + \overset{2}{n} \qquad b = \overset{3}{m}\,\overset{2}{n}$$
$$= 3 + 2 \qquad\quad = (3)(2)$$
$$= 5 \qquad\qquad\ \ = 6$$

Now,
$$W = \overset{5}{a}\,\overset{6}{b}$$
$$= (5)(6)$$
$$= 30$$

Sample Problem 10-2 Compute the value of J using the formulas $J = (a + b)/a$, $a = m^2$, and $b = (m + n)/2$, given the data $m = 3$ and $n = 5$.

Solution:
$$m = 3 \qquad\qquad n = 5$$
$$a = m^2 = 3^2 = 9 \qquad b = \frac{\overset{3}{m} + \overset{5}{n}}{2} = \frac{3 + 5}{2} = \frac{8}{2} = 4$$
$$J = \frac{\overset{9}{a} + \overset{4}{b}}{\underset{9}{a}} = \frac{9 + 4}{9} = \frac{13}{9} = 1\frac{4}{9}$$

Sample Problem 10-3

Compute the value of L using the formulas $L = a^2 + ab$, $a = 2m + 1$, $b = 2n - 3$, $m = rs$, and $n = r + s$, given the data $r = 2$ and $s = 1$.

Strategy:

In order to find L using $L = a^2 + ab$, the values of a and b must be known. In order to find a and b using $a = 2m + 1$ and $b = 2n - 3$, the values of m and n must be known. The values of m and n can be found by substituting $r = 2$ and $s = 1$ into the formulas $m = rs$ and $n = r + s$.

Solution:

$r = 2 \qquad s = 1$

$m = \overset{2}{r}\overset{1}{s} \qquad n = \overset{2}{r} + \overset{1}{s}$
$ = (2)(1) \qquad = 2 + 1$
$ = 2 \qquad = 3$

$a = 2\overset{2}{m} + 1 \qquad b = 2\overset{3}{n} - 3$
$ = 2(2) + 1 \qquad = 2(3) - 3$
$ = 4 + 1 \qquad = 6 - 3$
$ = 5 \qquad = 3$

$L = \overset{5}{a}{}^2 + \overset{5}{a}\overset{3}{b}$
$ = 5^2 + (5)(3)$
$ = 25 + 15$
$ = 40$

Working problems of the type above depends, to a great extent, on keeping your work well organized. How much scratch work you write above each formula depends on how much you can do in your head.

Sec. 10-2 Problems Involving More than One Formula

Complete the problems below. Completed problems appear on page 326.

Sample Set

(a) Compute the value of K using the formulas $K = a + b$, $a = 3m$, and $b = n^3$, given $m = 4$ and $n = 3$.

$a = 3m$ $b = n^3$ $K = a + b$

$= 3(\ \ )$ $= (\ \ )^3$ $= \boxed{} + \boxed{}$

$= \boxed{}$ $= \boxed{}$ $= \boxed{}$

(b) Compute the value of J using the formulas $J = (a+b)/c$, $a = m + 2$, $b = m + n$, $c = mn$, $m = r/s$, and $n = r - s$, given $r = 4$ and $s = 2$.

$m = \dfrac{r}{s} = \dfrac{\boxed{}}{\boxed{}} = \boxed{}$ $n = r - s$

$= \boxed{} - \boxed{} = \boxed{}$

$b = m + n$ $a = m + 2$ $c = mn$

$= \boxed{} + \boxed{}$ $= \boxed{} + 2$ $= (\ \ )(\ \ )$

$= \boxed{}$ $= \boxed{}$ $= \boxed{}$

$J = \dfrac{a+b}{c} = \dfrac{\boxed{} + \boxed{}}{\boxed{}} = \dfrac{\boxed{}}{\boxed{}} = \boxed{}$

Completed Problems

(a) Compute the value of K using the formulas $K = a + b$, $a = 3m$, and $b = n^3$, given $m = 4$ and $n = 3$.

$a = 3m \qquad\qquad b = n^3 \qquad\qquad K = a + b$

$= 3(\boxed{4}) \qquad\quad = (\boxed{3})^3 \qquad\quad = \boxed{12} + \boxed{27}$

$= \boxed{12} \qquad\qquad = \boxed{27} \qquad\qquad = \boxed{39}$

(b) Compute the value of J using the formulas $J = (a+b)/c$, $a = m + 2$, $b = m + n$, $c = mn$, $m = r/s$, and $n = r - s$, given $r = 4$ and $s = 2$.

$m = \dfrac{r}{s} = \dfrac{\boxed{4}}{\boxed{2}} = \boxed{2} \qquad n = r - s$

$\phantom{m = \dfrac{r}{s} = \dfrac{4}{2} = 2 \qquad n} = \boxed{4} - \boxed{2} = \boxed{2}$

$b = m + n \qquad\qquad a = m + 2 \qquad\qquad c = mn$

$= \boxed{2} + \boxed{2} \qquad\quad = \boxed{2} + 2 \qquad\quad = (\boxed{2})(\boxed{2})$

$= \boxed{4} \qquad\qquad\quad = \boxed{4} \qquad\qquad = \boxed{4}$

$J = \dfrac{a+b}{c} = \dfrac{\boxed{4} + \boxed{4}}{\boxed{4}} = \dfrac{\boxed{8}}{\boxed{4}} = \boxed{2}$

Name: _____

Class: _____

Exercises for Sec. 10-2   Find the value of K using the given formulas and data.

1. $K = a + b$ $a = \dfrac{m}{2}$, $b = mn$; $m = 4$, $n = 2$

2. $K = a - b$ $a = 2m$, $b = mn$; $m = 3$, $n = 5$

3. $K = 2a + b$ $a = mn$, $b = \dfrac{m + n}{2}$; $m = 3$, $n = 5$

4. $K = 3a - b$ $a = \dfrac{m}{n}$, $b = m - n$; $m = 6$, $n = 2$

5. $K = a(a + b)$ $a = \dfrac{m}{n}$, $b = m^2$; $m = 4$, $n = 2$

6. $K = a(a - b)$ $a = m^3$, $b = mn$; $m = 3$, $n = 2$

7. $K = \dfrac{a+b}{a}$ $a = 3m$, $b = m+n$; $m = 2$, $n = 4$

8. $K = \dfrac{a-b}{b}$ $a = 2n$, $b = m-n$; $m = 4$, $n = 2$

9. $K = am - b$ $a = m^3$, $b = mn$; $m = 2$, $n = 4$

10. $K = a - bn$ $a = \dfrac{m}{n}$, $b = m^2$; $m = 4$, $n = 2$

11. $K = a - b$ $a = m+n$, $b = mn$, $m = \dfrac{r}{2}$, $n = rs$; $r = 4$, $s = 2$

12. $K = a + b$ $a = mn$, $b = m+n$, $m = 2r$, $n = rs$; $r = 3$, $s = 2$

13. $K = \dfrac{a+b}{a}$ $a = 3m$, $b = 2n+1$, $m = r+s$, $n = r-s$; $r = 4$, $s = 2$

14. $K = \dfrac{a-b}{a}$ $a = 2m$, $b = 2n - 1$, $m = rs$, $n = \dfrac{r}{s}$; $r = 4$, $s = 2$

15. $K = (a-b)^2$ $a = m^2$, $b = \dfrac{n}{m}$, $m = \dfrac{r}{s}$, $n = rs$; $r = 4$, $s = 2$

16. $K = (a+b)^2$ $a = mn$, $b = n^2$, $m = r - s$, $n = \dfrac{r}{s}$; $r = 6$, $s = 2$

17. $K = \dfrac{a+b}{ab}$ $a = m + n$, $b = m - n$, $m = rs$, $n = \dfrac{r}{s}$; $r = 4$, $s = 2$

18. $K = \dfrac{ab}{a-b}$ $a = m - n$, $b = m + n$, $m = r + s$, $n = r - s$; $r = 6$, $s = 2$

19. $K = a^3 + ab^2$ $a = m - n$, $b = \dfrac{m}{n}$, $m = r + s$, $n = 2r$; $r = 2$, $s = 6$

20. $K = a^2 + ab^3 \quad a = \dfrac{n}{m}, \quad b = m + n, \quad m = r - s, \quad n = 2r; \quad r = 4, \quad s = 3$

21. $K = a + b + c \quad a = m + n, \quad b = m - n, \quad c = 2n, \quad m = r - s,$
    $n = r + s; \quad r = 6, \quad s = 2$

22. $K = a + b - c \quad a = m - n, \quad b = m + n, \quad c = 2m, \quad m = r + s,$
    $n = r - s; \quad r = 5, \quad s = 2$

23. $K = a(b + c) \quad a = \dfrac{m}{n}, \quad b = 2m, \quad c = m - n, \quad m = 2r, \quad n = \dfrac{s}{3};$
    $r = 2, \quad s = 3$

24. $K = a(b - c) \quad a = \dfrac{n}{m}, \quad b = \dfrac{m}{2}, \quad c = m + n, \quad m = \dfrac{r}{2}, \quad n = 3s;$
    $r = 4, \quad s = 2$

25. $K = \dfrac{a + b}{c} \quad a = mn, \quad b = m + n, \quad c = 2m, \quad m = r + s - t, \quad n = \dfrac{rs}{t};$
    $r = 4, \quad s = 3, \quad t = 6$

26. $K = \dfrac{a-b}{c}$ $a = m - n$, $b = 3n$, $c = \dfrac{m}{2}$; $m = r + s + t$, $n = rs - t$; $r = 4$, $s = 3$, $t = 5$

27. $K = a + b$ $a = mn$, $b = \dfrac{m}{n}$, $m = r + s$, $n = r - s$, $r = u + v$, $s = u - v$; $u = 4$, $v = 2$

28. $K = a + b$ $a = m^2$, $b = \dfrac{n+3}{m}$, $m = r - s$, $n = r + s$, $r = uv$, $s = \dfrac{u}{v}$; $u = 6$, $v = 2$

29. $K = \dfrac{a}{b}$ $a = m - n$, $b = m^2$, $m = r - 2$, $n = r + s$, $r = uv$, $s = \dfrac{u}{v}$; $u = 6$, $v = 3$

30. $K = \dfrac{a}{b}$ $a = m - n$, $b = mn$, $m = r + s$, $n = r - s$, $r = u + v$, $s = u - v$; $u = 4$, $v = 2$

Section 10-3 Symbolic Representation

Several phrases are written below. The word or words in each phrase indicate a mathematical operation. An equivalent phrase can be written using mathematical symbols.

Verbal	Symbolic
A plus B	$A + B$
A minus B	$A - B$
A times B	$A \times B$ or AB
A divided by B	$A \div B$ or $\dfrac{A}{B}$

Examples

(a) Verbal: the product of A and B
Symbolic: $A \times B$ or AB
— The word *product* indicates that A and B are multiplied.

(b) Verbal: the sum of A and B
Symbolic: $A + B$
— The word *sum* indicates that A and B are added.

(c) Verbal: the difference when A is subtracted from B
Symbolic: $B - A$
— To find the difference, A must be subtracted **from** B.

(d) Verbal: 5 more than A
Symbolic: $A + 5$
— In order to find 5 more than A, add 5 to A.

(e) Verbal: 3 less than B
Symbolic: $B - 3$
— In order to find 3 less than B, subtract 3 **from** B.

(f) Verbal: D times 7
Symbolic: $7D$
— It is conventional to place the number in front of the letter (not $D7$).

More than one operation:

(g) Verbal: A times B plus C
Symbolic: $AB + C$
— Multiplication before addition.

(h) Verbal: A times the sum of B and C
Symbolic: $A(B + C)$
— A is multiplied by a sum; this means that B and C must first be added; parentheses are used to indicate this.

(i) Verbal: A minus B squared
Symbolic: $A - B^2$
— Squaring is a multiplication operation; multiplication precedes subtraction; same as A minus the square of B.

(j) Verbal: A times B, that product squared
Symbolic: $(AB)^2$
— A product is being squared; A and B must first be multiplied; parentheses are used to indicate this.

334 Formulas

Using the word *quantity*

(k) Verbal: *A* minus *B*, that quantity squared
Symbolic: $(A - B)^2$

The word *quantity* indicates the subtraction is to be done first; that difference is then squared.

(l) Verbal: *A* plus *B*, that quantity divided by *C*
Symbolic: $(A + B) \div C$, or better,
$$\frac{A + B}{C}$$

The word *quantity* indicates the addition is to be performed before the division.

Complete each symbolic phrase below. Completed phrases appear on page 335.

Sample Set

(a) *A* more than *B*   (b) *A* less than *B*   (c) *A* less *B*

☐ + ☐   ☐ − ☐   ☐ − ☐

(d) 5 more than *A* times *B*   (e) 3 times the sum of *A* and *B*

☐☐ + ☐   ☐(☐ + ☐)

(f) *A* plus the quantity *B* divided by *C*

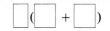

(g) *A* minus *B*, that difference divided by *C*

(h) *A* plus *B*, the square of that sum divided by the product of *A* and *B*

Completed Phrases

(a) A more than B (b) A less than B (c) A less B

$\boxed{B} + \boxed{A}$ $\boxed{B} - \boxed{A}$ $\boxed{A} - \boxed{B}$

(d) 5 more than A times B   (e) 3 times the sum of A and B

$\boxed{A}\boxed{B} + \boxed{5}$ $\boxed{3}(\boxed{A} + \boxed{B})$

(f) A plus the quantity B divided by C

$\boxed{A} + \dfrac{\boxed{B}}{\boxed{C}}$

(g) A minus B, that difference divided by C

$\dfrac{\boxed{A} - \boxed{B}}{\boxed{C}}$

(h) A plus B, the square of that sum divided by the product of A and B

$\dfrac{(\boxed{A} + \boxed{B})^2}{\boxed{A}\boxed{B}}$

Name: _____

Class: _____

Exercises for Sec. 10-3 Represent each phrase using mathematical symbols.

1. A plus 5
2. 5 more than A
3. The sum of 5 and A
4. A minus 5
5. 5 less than A
6. A less 5
7. B added to A
8. B subtracted from A
9. A times B
10. The product of A and B
11. A divided by B
12. The sum when A is added to B
13. The difference when A is subtracted from C
14. The square of A plus B
15. A squared times B
16. A squared divided by B
17. The square of A minus B
18. The square of the product of A and B
19. The square of the quotient when A is divided by B
20. A plus B, that sum cubed
21. A minus B, that difference cubed
22. A times the sum of B and C

23. A times the difference B minus C

24. L divided by K, that quotient times J.

25. L times K, that product divided by J

26. L times the quotient K divided by J

27. L divided by the product K times J

28. L less than K times J

29. L more than K divided by J

30. 10 more than the square of the sum of A and B

31. 5 less than the cube of the difference A minus B

32. 10 less than A cubed minus B

33. 5 more than A squared plus B

34. R plus S times the quotient T divided by W

35. R minus the quantity W times the quotient S divided by T

36. 6 more than A times B, that sum divided by the sum of C and D

37. 6 less than A times B, that difference divided by the difference A minus B

38. *M* plus 5 times *L* minus three *N*, that quantity divided by *S*

39. *A* plus *B* times *C*, that quantity divided by the product *R* times *S* times *T*

40. *A* plus *B* squared, the cube of that sum divided by the square of the difference *A* squared minus *B*

Section 10-4 Mathematical Statements

A mathematical statement expresses a relationship between quantities. Examples of statements of equality are given below.

$$3x = 9$$
$$A = lw$$
$$\text{Area} = \text{length} \times \text{width}$$
Area is the product of length and width

In this section, a formula of the form $A = lw$ will be referred to as a **symbolic formula** (all letters). A formula of the form *area = length × width* will be called a **verbal formula.**

The object in this lesson will be to write symbolic or verbal formulas equivalent to mathematical statements that are given in the form of sentences. In many instances, a verbal formula is preferable to a symbolic formula. This is true in cases where letters are not normally used to represent the quantities involved, or when the user is not accustomed to working with symbolic formulas.

Sample Problem 10-4

Write a symbolic formula that represents the statement:

L is A plus B, that sum divided by K

Solution:

$$L = (A + B) \div K$$

or

$$L = \frac{A + B}{K}$$

The words *that sum* indicate that the addition is to be done before the division. This is indicated by using parentheses in the formula.

Sample Problem 10-5

Write a symbolic formula that represents the statement:

Volume is the product of the length, width, and height

Solution:

$$V = lwh$$

The letters V, l, w, and h are commonly used to represent volume, length, width, and height.

Important Point

In the sample problem just completed, it is important to realize that the letters, V, l, w, and h are not initials representing words, but rather letters representing numbers.

Sample Problem 10-6

Write a verbal formula that represents the statement:

The **selling price is** the sum of the **cost** of the item **plus** the **margin**

Solution:

$$\text{Selling price} = \text{cost} + \text{margin}$$

The words that are essential to writing the formula are in bold type. Including the words *of the item* would make the formula clumsy. It might be assumed that the item was mentioned in an earlier sentence, and that it is understood what the cost refers to. It would not be correct to write

$$\text{Price} = \text{cost} + \text{margin} \quad \leftarrow \text{incorrect}$$

One would not know whether *Price* referred to selling price or list price. They may not be the same.

Sample Problem: 10-7

Write a verbal formula that represents the statement:

Joe's age is four more than Jim's age

Solution:

Joe's age = Jim's age + 4

Since Joe's age is four more than Jim's age, 4 must be added to Jim's age to obtain Joe's age.

Can you see what would be wrong in writing

Joe = Jim + 4

It would not be clear if the relationship is age, weight, height, or some other quantity relating to Joe and Jim.

Sample Problem: 10-8

Write a verbal formula that represents the statement:

In baseball, the **total** number of **bases** credited to a batter **is one times** the number of **singles plus two times** the number of **doubles plus three times** the number of **triples plus four times** the number of **home runs** he attains

Solution:

Total bases = 1 × no. singles + 2 × no. doubles
　　　　　　　+ 3 × no. triples + 4 × no. home runs

The modifying phrases "in baseball," "credited to a batter," and "he attains" are informative, but should not be included in the formula. It may be misleading to write

Total bases = 1 singles + 2 doubles
　　　　　　　+ 3 triples + 4 home runs

For example, 2 doubles might be interpreted as the number of doubles; that is, two of them, rather than 2 times the number of doubles.

The last sample problem demonstrates the value of writing a verbal formula. An exceedingly long sentence was replaced by a short verbal formula. Looking at the verbal formula, one can see at once the relationship between the quantities involved.

Complete the following formulas. Completed formulas appear on page 344.

Sample Set

In problems *a* and *b*, write a symbolic formula that represents each statement.

(a) *L* is five times the sum of *A* and *B*.

$$\square = \square\,(\square + \square\,)$$

(b) *K* is *A* plus *B*, that sum divided by the cube of the product of *C* and *D*.

$$K = \frac{\square + \square}{(\square\,\square\,)^3}$$

In problems *c* to *e*, write a verbal formula that represents each statement.

(c) John's weight is 5 pounds less than Jim's weight.

_____ = _____ − _____

(d) The commission is 10 percent of the selling price.

_____ = _____ × _____

(e) The area is one-half the altitude times the sum of the upper base and the lower base.

_____ = _____ _____

(_____ + _____)

Completed Formulas

(a) L is five times the sum of A and B.

$$\boxed{L} = \boxed{5} \, (\boxed{A} + \boxed{B})$$

(b) K is A plus B, that sum divided by the cube of the product of C and D.

$$K = \frac{\boxed{A} + \boxed{B}}{(\boxed{C}\;\boxed{D})^3}$$

(c) John's weight is 5 pounds less than Jim's weight.

$$\underline{\text{John's weight}} = \underline{\text{Jim's weight}} - \underline{5}$$

(d) The commission is 10 percent of the selling price.

$$\underline{\text{Commission}} = \underline{10\%} \times \underline{\text{selling price}}$$

(e) The area is one-half the altitude times the sum of the upper base and the lower base.

$$\underline{\text{area}} = \underline{\tfrac{1}{2}} \; \underline{\text{altitude}} \; (\underline{\text{upper base}} + \underline{\text{lower base}})$$

Name: _____

Class: _____

Exercises for Sec. 10-4 Write a symbolic formula that represents each statement.

1. R is the sum of P and Q.

2. J is 5 less than A.

3. Q is twice the difference A minus B.

4. L is R times S plus 5.

5. M is A times the sum of three times B and C.

6. L is A minus B, that difference divided by C.

7. J is 6 more than the product of M and N.

8. K is A divided by B plus C.

9. S is 5 less than the sum of M and N.

10. Q is 2 less than the square of B.

11. D is one-half the sum, A plus the square of C.

12. R is the square of the sum of L plus J.

13. N is the sum of A plus B, that sum divided by the difference A minus B.

14. L is M times the square of R.

15. V is four-thirds π times the cube of R.

16. The area is π times the square of the radius, where A = area and r = radius.

17. Density is weight divided by volume, where d = density, W = weight, and V = volume.

18. Force is mass times acceleration, where F = force, m = mass, and a = acceleration.

19. The number of volts is the product of the number of amps times the number of ohms, where E = number of volts, I = number of amps, and R = number of ohms.

20. The number of watts is the product of the resistance and the square of the amps, where P = number of watts, R = number of ohms of resistance, and I = number of amps.

21. Interest is principal times rate times time, where i = interest, p = principal, r = rate, and t = time.

22. The expansion of a metal beam due to a rise in temperature is the the product of the length of the beam times the temperature change times a constant K. K depends only on the material of which the beam is made, and E = expansion, L = length, ΔT = temperature change, and K = constant.

23. Perimeter is twice the length added to twice the width, where P = perimeter, l = length, and w = width.

Write a verbal formula that represents each statement.

24. Bob's age is 5 more than Jim's age.

25. Mary is twice as old as Alice.

26. Ron is 20 years younger than George.

27. Carol's age is the sum of Susan's age and Anne's age.

28. The margin is the sum of profit and overhead.

29. Cost is selling price minus margin.

30. Profit is 15 percent of list price.

31. Selling price is list price minus the markdown.

32. Percent of profit is 100 times the quotient of profit divided by selling price.

33. The sales tax is 6 percent of the selling price.

34. The discount is 20 percent of the list price.

35. The distance traveled is the average speed multiplied by the number of hours one drives.

36. The number of miles per gallon is the number of miles driven divided by the number of gallons used.

37. The price per pound is the price divided by the number of pounds.

38. Net income is gross income minus deductions.

39. The wage is the hourly rate times the hours worked.

40. To find the intelligence quotient (IQ), we divide the mental age by the chronological age and then multiply that quotient by 100.

Section 10-5 Problems Involving More than One Verbal Formula

Texts in many areas will state mathematical relationships between quantities in the form of sentences. Quite often, several of these sentences are given, forming a paragraph. The student who can replace these statements with verbal formulas and work with these formulas has a great advantage. Such a student can progress independently and does not have to memorize procedures given by an instructor.

Sample Problem 10-9

Joe is 5 years older than Jim. Jim's age is the sum of Bob's age and Don's age. Bob's age is twice Mary's age.

Data: Mary is 2 years old, and Don is 5 years old.

Compute Joe's age.

Solution:

Step 1. Write a verbal formula for each sentence; leave space between the formulas

Joe's age = Jim's age + 5

Jim's age = Bob's age + Don's age

Bob's age = 2 × Mary's age

Step 2. Substitute the data into the formulas and compute all values

$$\underbrace{\text{Joe's age}}_{14} = \underbrace{\text{Jim's age}}_{9} + 5$$

$$\underbrace{\text{Jim's age}}_{9} = \underbrace{\text{Bob's age}}_{4} + \underbrace{\text{Don's age}}_{5}$$

$$\underbrace{\text{Bob's age}}_{4} = 2 \times \underbrace{\text{Mary's age}}_{2}$$

Conclusion: Joe is 14 years old.

Sample Problem 10-10

The **assessed value** of a home **is one-third of** the **market value**. The **number of hundreds** of assessed value **is the assessed value divided by $100**. The yearly **property tax is** the **number of hundreds** of assessed value **multiplied by** the **tax rate** per $100 of assessed value.

Data: Market value = $60,000, tax rate = $7.50 per $100.

Compute the property tax.

350 Formulas

Solution:

$$\underbrace{\text{Assessed value}}_{\$20{,}000} = \tfrac{1}{3} \times \underbrace{\text{market value}}_{\$60{,}000}$$

$$\underbrace{\text{Number of hundreds}}_{200} = \frac{\overbrace{\text{assessed value}}^{\$20{,}000}}{\$100}$$

$$\underbrace{\text{Property tax}}_{\$1{,}500} = \underbrace{\text{number of hundreds}}_{200} \times \underbrace{\text{tax rate}}_{\$7.50}$$

Conclusion: The property tax will be $1,500.

In Sample Problem 10-10, the words that are essential to writing verbal formulas are printed in bold type. In the problems you will work, you must select the key words.

Sometimes a sentence will be included in a paragraph that provides information, but does not state a mathematical relationship between quantities. No verbal formula represents such a statement. An example is provided by the third sentence in the following problem.

Sample Problem 10-11

The **selling price** of an item **is** the sum of the **cost** of the item to the dealer **plus** his **margin**. **Margin is** defined to be the **overhead**, such as rent, taxes, salaries, and commission, **plus** the **profit** he receives. Occasionally the selling price is not the original list price of the item, owing to the fact that the list price has been marked down. In this case, the **selling price is** the **list price minus** the amount the item is **marked down**.

Data: List price = $200, markdown = $30, overhead = $40, cost = $100.

Compute the profit.

Solution: Step 1. Write verbal formulas, and substitute in the data

$$\text{Selling price} = \underbrace{\text{cost}}_{\$100} + \text{margin}$$

$$\text{Margin} = \underbrace{\text{overhead}}_{\$40} + \text{profit}$$

$$\text{Selling price} = \underbrace{\text{list price}}_{\$200} - \underbrace{\text{markdown}}_{\$30}$$

Step 2. The third formula tells us that selling price = $170; substituting $170 into the first formula, a simple equation is solved mentally to find margin = $70; substituting $70 into the second equation, another equation is solved to find profit = $30

$$\underbrace{\text{Selling price}}_{\$170} = \underbrace{\text{cost}}_{\$100} + \underbrace{\text{margin}}_{\$70}$$

$$\underbrace{\text{Margin}}_{\$70} = \underbrace{\text{overhead}}_{\$40} + \underbrace{\text{profit}}_{\$30}$$

$$\underbrace{\text{Selling price}}_{\$170} = \underbrace{\text{list price}}_{\$200} - \underbrace{\text{markdown}}_{\$30}$$

Name: _____

Class: _____

Exercises for Sec. 10-5   Information for Exercises 1 to 3: Bob's age is Bill's age subtracted from John's age. Bill is 4 years older than Jim. John's age is Alice's age divided by 3.

1. Data: Alice's age = 30; Jim's age = 3.
   Compute Bob's age.

2. Data: Bob's age = 17; Jim's age = 8.
   Compute Alice's age.

3. Data: Bob's age = 5; Alice's age = 45.
   Compute Jim's age.

Information for Exercises 4 to 6: Mary is 5 pounds lighter than Helen. Ralph's weight is the sum of Mary's weight and Alice's weight. Ralph eats a lot of candy. Alice is 10 pounds heavier than John.

4. Data: John's weight = 80; Ralph's weight = 200.
   Compute Helen's weight.

5. Data: John's weight = 90; Helen's weight = 120.
   Compute Ralph's weight.

6. Data: Ralph's weight = 250; Helen's weight = 125.
   Compute John's weight.

Information for Exercises 7 and 8: A person's net income is his gross income minus deductions. Assume that the total of the deductions is the sum of taxes, pension deductions, and insurance that the person pays.

7. Data: Tax = $80; pension = $30; insurance = $15; gross income = $450.
   Compute the net income.

8. Data: Tax = $160; pension = $80; insurance = $20; net income = $690.
   Compute the gross income.

Information for Exercises 9 and 10: The selling price of an item is the sum of the cost of the item plus the margin. The margin is the sum of the profit and the overhead.

9. Data: Profit = $20; overhead = $50; selling price = $280.
   Compute the cost.

10. Data: Selling price = $180; cost = $120; overhead = $20.
    Compute the profit.

Information for Exercises 11 and 12: A farmer's gross income from a corn harvest is the total number of bushels harvested times the price per bushel. The total number of bushels harvested is the average number of bushels per acre times the number of acres he has planted in corn.

11. Data: Gross income = $40,000; price per bushel = $2; average bushels per acre = 100.
    Compute the number of acres planted.

12. Data: Gross income = $24,000; acres planted = 100; average bushels per acre = 80.
    Compute the price per bushel.

Information for Exercises 13 and 14: The gas mileage of a car is the quotient obtained when the distance driven is divided by the number of gallons used. The distance driven is the product when the rate of speed is multiplied by the time the car is driven.

13. Data: Speed = 50; time = 6; gallons = 20.
    Compute the gas mileage.

14. Data: Gallons = 10; gas mileage = 18; time = 3.
    Compute the speed.

Information for Exercises 15 and 16: The total cost of an item is the cost of the item to the dealer (what she pays for it) added to her overhead. The profit the dealer makes is the total cost subtracted from the selling price. Her percent of profit is 100 times the profit divided by the total cost.

15. Data: Cost (to the dealer) = $150; overhead = $50; percent profit = 20. Compute the selling price.

16. Data: Cost (of item) = $100; overhead = $50; selling price = $180.
    Compute the percent of profit.

17. The total owed a loan company is the amount borrowed plus the interest. The interest is stated as the interest rate times the amount borrowed times the number of years of the loan. The number of months of the loan is 12 times the number of years of the loan. The monthly payment is the total owed divided by the number of months of the loan.
    Data: Amount borrowed = $2,000; interest rate = 8 percent; number of years = 5.
    Compute the monthly payment.

18. The density of an object is defined to be the weight of the object divided by its volume. The volume of a gold brick is the product of its length, width, and height. The value of the gold brick is equal to its weight (in ounces) times the cost of gold per ounce.
    Data: Length = 6 inches; width = 3 inches; height = 2 inches; density = 11.53 ounces per cubic inch; cost per ounce = $135.
    Compute the value of the brick.

Information for Exercises 19 to 22: The number of watts of electrical power is the product of the number of volts of electromotive force multiplied by the number of amps of current in a circuit. The number of volts is the product of the number of amps times the number of ohms of resistance in the circuit.

In common usage, the number of watts is represented by the letter P, number of volts by E, number of amps by I, and the resistance by R.

19. Data: $I = 15$ amps; $R = 14$ ohms.
    Compute P.

20. Data: $P = 1,000$ watts; $E = 220$ volts.
    Compute R.

21. Data: $P = 550$ watts; $I = 5$ amps.
    Compute R.

22. Data: $E = 200$ volts; $R = 40$ ohms.
    Compute P.

23. In baseball, the batting average of a player is defined to be the number of hits he makes divided by the number of times he is at bat. The number of hits he makes is the sum of the number of singles, doubles, triples, and home runs he makes. The number of times that he is credited with being at bat is the number of hits he gets added to the number of outs he makes.
    Data: Singles = 80; doubles = 20; triples = 5; home runs = 20; outs = 275.
    Compute the batting average (round to three places).

24. A baseball batter's slugging average is defined to be the total bases he attains divided by the number of times he is at bat. The total bases he attains is 1 times the number of singles plus 2 times the number of doubles plus 3 times the number of triples plus 4 times the number of home runs he gets. The number of times he is at bat is the sum of the number of singles, doubles, triples, and home runs he gets plus the number of outs he makes.
    Data: Singles = 80; doubles = 20; triples = 5; home runs = 20; outs = 275.
    Compute the slugging average (round to three places).

25. To find a pitcher's earned-run average (era), one first multiplies the number of earned runs he gives up by 9. That product is then divided by the total number of innings he pitches.
    Data: Number of innings pitched = 198; number of earned runs = 77.
    Compute the earned-run average (round to two decimal places).

Name: _____

Class: _____

Unit 10 Test Find the value of J in each formula.

1. $J = 3a + 2b$ $a = 3, b = 4$

2. $J = ab^2 + 2$ $a = 2, b = 3$

3. $J = a(a - b)$ $a = 3, b = 5$

4. $J = \dfrac{(a + b)^2}{ab}$ $a = 3, b = 2$

Find the value of J in each set of formulas.

5. $J = ab$ $a = m + n,\ b = \dfrac{m}{n};\ m = 6,\ n = 2$

6. $J = \dfrac{a - b}{a}$ $a = 2m - n,\ b = 2m + 3n;\ m = 3,\ n = 2$

7. $J = \dfrac{a}{b}$ $a = m + n$, $b = 2m$, $m = \dfrac{r}{s}$, $n = rs$; $r = 4$, $s = 2$

8. $J = (a + b)^2$ $a = m - 3n$, $b = \dfrac{m}{n}$, $m = r + s$, $n = r - s$; $r = 5$, $s = 3$

Represent each phrase using mathematical symbols.

9. Twice the sum of a and b.

10. 5 less than m times n

11. r minus s, that quantity cubed

12. c times d, that product divided by the sum of c and d

Write a symbolic formula representing each statement.

13. K is the sum of a and b.

14. L is m times the square of n.

15. M is r plus s, that quantity divided by t.

16. V is a times the square of the sum of a and b.

Write a verbal formula representing each statement in Probs. 17 and 18.

17. Bob's age is 4 years more than twice Jim's age.

18. A halfback's running average is the yards he gains minus the yards he loses, that difference divided by the number of times he carries the ball.

19. John's age is Bill's age minus Richard's age. Richard's age is Mary's age divided by Nancy's age. Mary's age is the sum of Nancy's age plus David's age.
    Data: Nancy's age = 2 years; David's age = 8 years; Bill's age = 15 years.
    Compute John's age.

20. The markdown is 20 percent of the list price. Selling price is the list price minus the markdown. Selling price is also the sum of margin plus cost.
    Data: Margin = $20; list price = $100.
    Compute the cost.

UNIT 11
DENOMINATE NUMBERS

Section 11-1 Conversion Factors

A **denominate number** is a quantity such as 9 feet in which a **unit** or denomination of measure is given. In the denominate number 9 feet, the unit of measure is feet. Other examples of denominate numbers are 81.2 pounds, 15 amps, $12.50, 215 square miles, and 327 people.

Conversion Factors Two denominate numbers, with different units of measure, may be equivalent. For example,

$$1 \text{ mile} = 1{,}760 \text{ yards}$$

Each of these denominate numbers represents the same length.

When equivalent denominate numbers form the numerator and denominator of a fraction, the fraction is equivalent to 1.

$$\frac{1{,}760 \text{ yards}}{1 \text{ mile}} = \frac{1 \text{ mile}}{1 \text{ mile}} = 1$$

When such a fraction is used as a multiplier, it is called a **conversion factor**.

Sample Problem 11-1

2 miles is how many yards?

Strategy: In this problem, 2 miles is to be converted into yards. You must multiply 2 miles by a conversion factor that cancels miles and introduces yards into the problem.

Solution:
$$2 \text{ mi} = 2 \text{ mi} \times 1$$
$$= \frac{2 \cancel{\text{mi}}}{1} \times \frac{1{,}760 \text{ yd}}{1 \cancel{\text{mi}}}$$
$$= 3{,}520 \text{ yd}$$

The procedure used in the sample problem is similar to multiplying a whole number by 1. For example,

$$2 = 2 \times 1$$
$$= 2 \times \tfrac{3}{3}$$
$$= \tfrac{6}{3}$$

The numbers 2 and $\tfrac{6}{3}$ are equivalent since the conversion factor $\tfrac{3}{3}$ equals 1. In the same way, 2 miles and 3,520 yards are equivalent since the conversion factor 1,760 yards/1 mile equals 1.

The conversion factor must be chosen so that the original units are canceled and the desired units are introduced.

Rates If you travel 200 miles in 4 hours, what would be your average speed?

$$\text{Answer} = \frac{\overset{50}{\cancel{200}}\text{ mi}}{\underset{1}{\cancel{4}}\text{ hr}} = \frac{50 \text{ mi}}{1 \text{ hr}} = 50 \text{ mi/hr} \quad \text{read as 50 mi per hr}$$

The word *per* is used to indicate a **rate**. Note that it also indicates a division has taken place; in this case, number of miles divided by number of hours.

Further examples of rates:

1 If you buy 8 pounds of apples for 108 cents, you are buying apples at a rate of $13\frac{1}{2}$ cents per pound.

$$\frac{\overset{13\frac{1}{2}}{\cancel{108}}\cancel{c}}{\underset{1}{\cancel{8}}\text{ lb}} = \frac{13\frac{1}{2}\cancel{c}}{1 \text{ lb}} = 13\frac{1}{2}\cancel{c}/\text{lb}$$

2 If you earn $30 for working 6 hours, you are earning money at a rate of $5 per hour.

$$\frac{\overset{5}{\cancel{\$30}}}{\underset{1}{\cancel{6}}\text{ hr}} = \frac{\$5}{1 \text{ hr}} = \$5/\text{hr}$$

3 Yards "occur" in a mile at a rate of 1,760 yards per mile.

$$\frac{1{,}760 \text{ yd}}{1 \text{ mi}} = 1{,}760 \text{ yd/mi}$$

Constant Rates as Conversion Factors If you buy 8 pounds of apples for 108 cents ($13\frac{1}{2}$¢ per pound), can you expect to buy 16 pounds of apples for 216 cents (still $13\frac{1}{2}$¢ per pound)? Quite often, larger amounts are sold at a cheaper rate. If the rate remains constant (the same), the quantities pounds of apples and price of apples are said to be **directly proportional**. This will mean that twice as many pounds will cost twice as much, one-third as many pounds will cost one-third as much, and so on.

In the problems throughout this section, it will be assumed that rates remain constant. Each rate may then be used as a conversion factor.

Sample Problem 11-2

If 8 pounds of apples cost 108 cents, how much will 14 pounds cost? (*Note:* The apples are being sold at the rate of 108 cents per 8 pounds.)

Solution:

Method 1. Find the price of 1 pound of apples:

$$\frac{108¢}{8 \text{ lb}} = \frac{13\frac{1}{2}¢}{1 \text{ lb}} \qquad 108 \div 8 = 13\frac{1}{2}$$

The cost of 14 lb can now be found using the formula

$$\text{Cost} = \text{number of pounds} \times \text{price per 1 lb}$$

$$\text{Cost} = 14 \text{ lb} \times \frac{13\frac{1}{2}¢}{1 \text{ lb}} = 189¢$$

Method 2. Without first reducing the rate:

$$\text{Cost} = \overset{7}{\cancel{14}} \cancel{\text{lb}} \times \frac{\overset{27}{\cancel{108¢}}}{\underset{1}{\underset{4}{\cancel{8\text{ lb}}}}} = 189¢$$

Note that the cancellation used in the second method simplifies the division and multiplication operations.

Sample Problem 11-3

If 8 pounds of apples cost 108 cents, how many pounds can be purchased with 162 cents? (*Note:* The apples are being sold at the rate of 8 pounds per 108 cents.)

Solution:

Method 1. Find the number of pounds that can be purchased for 1 cent:

$$\frac{\overset{2}{\cancel{8}} \text{ lb}}{\underset{27}{\cancel{108¢}}} = \frac{2 \text{ lb}}{27¢} = \frac{\frac{2}{27} \text{ lb}}{1¢} \qquad 2 \div 27 = \frac{2}{27}$$

The number of pounds that can be purchased can now be found using the formula

$$\text{Number of pounds} = \text{cost} \times \text{number of pounds per 1¢}$$

$$\text{Number of pounds} = \overset{6}{\cancel{162¢}} \times \frac{\frac{2}{27} \text{ lb}}{\cancel{1¢}} = 12 \text{ lb}$$

Method 2. Without first reducing the rate:

$$\text{Number of pounds} = \overset{6}{\cancel{162¢}} \times \frac{\overset{2}{\cancel{8}} \text{ lb}}{\underset{1}{\underset{27}{\cancel{108¢}}}} = 12 \text{ lb}$$

Sample Problem 11-4

It is found that a car can travel 375 miles on a 20-gallon tank of gasoline. If gasoline costs 60 cents per gallon, how much will it cost to drive the car 500 miles?

Solution:

Method 1. This method has two steps. First, find the number of gallons required to travel 500 mi using the rate 20 gal per 375 mi.

$$? \text{ gal} = \overset{4}{\cancel{500}} \cancel{\text{mi}} \times \frac{20 \text{ gal}}{\underset{3}{\cancel{375}} \cancel{\text{mi}}}$$

$$= \frac{80 \text{ gal}}{3} = 26\frac{2}{3} \text{ gal}$$

Second, find the cost of $26\frac{2}{3}$ gal using the rate 60¢/1 gal.

$$?¢ = 26\frac{2}{3} \cancel{\text{gal}} \times \frac{60¢}{1 \cancel{\text{gal}}} = 1{,}600¢ = \$16$$

Method 2. This method has one step. Note that, in the first method, 500 mi was first multiplied by 20 gal/375 mi. This product was then multiplied by 60¢/1 gal. Below, both conversion factors are used in a single product.

$$?¢ = \overset{4}{\cancel{500}} \cancel{\text{mi}} \times \frac{20 \text{ gal}}{\underset{\underset{1}{3}}{\cancel{375} \cancel{\text{mi}}}} \times \frac{\overset{20}{\cancel{60¢}}}{1 \cancel{\text{gal}}}$$

$$= 1{,}600¢ = \$16$$

Sample Problem 11-5

A certain kind of bolt is sold in boxes containing 50 bolts per box. If the cost is \$5 per 3 boxes, how much should 60 bolts cost?

Strategy:

In order to find the cost of 60 bolts, 60 bolts must be multiplied by conversion factors that will cancel "bolts" and introduce the desired unit, dollars. This can be accomplished by multiplying by

$$\frac{1 \text{ box}}{50 \text{ bolts}} \quad \text{not} \quad \frac{50 \text{ bolts}}{\text{box}}$$

and

$$\frac{\$5}{3 \text{ boxes}}$$

Solution:

$$\overset{\overset{2}{\cancel{20}}}{\cancel{60} \cancel{\text{bolts}}} \times \frac{1 \cancel{\text{box}}}{\underset{1}{\underset{\cancel{5}}{\cancel{50} \cancel{\text{bolts}}}}} \times \frac{\overset{1}{\cancel{\$5}}}{\underset{1}{\cancel{3} \cancel{\text{boxes}}}} = \$2$$

369 Sec. 11-1 Conversion Factors

It is true that the problem just worked could have been done by other methods. However, there are two great advantages to working problems using conversion factors.

1. Numbers may be canceled, making the arithmetic easy.
2. **If unwanted units cancel, and the desired unit is introduced, the problem has been set up correctly!**

In other words, the method is both **convenient** and **safe**.

Complete the problems below. Completed problems appear on page 370.

Sample Set

(a) If 12 pounds of bananas cost 63 cents, how much will 20 pounds cost?

$$20\,\text{lb} \times \frac{\Box\,¢}{\Box\,\text{lb}} = \Box\,¢$$

(b) If 15 percent of a number is 63, what is 25 percent of the number?

$$25\% \times \frac{\Box}{\Box\,\%} = \Box$$

(c) A salesperson sells an average of 63 pairs of shoes every 6 days. If the shoes sell for $20 per pair, what is the expected total dollar value of sales over a 100-day period?

$$100\,\text{days} \times \frac{\Box\,\text{pairs}}{\Box\,\text{days}} \times \frac{\$\,\Box}{1\,\text{pair}} = \$\,\Box$$

(d) If there are 6 ounces of meat in 15 ounces of spaghetti, and meat costs 3 cents per ounce, what is the cost of the meat in 100 ounces of spaghetti?

$$100\,\text{oz spaghetti} \times \frac{\Box\,\text{oz meat}}{\Box\,\text{oz spaghetti}} \times \frac{\Box\,¢}{\Box\,\text{oz meat}} = \Box\,¢$$

(e) A motorist can drive his car 15 miles on 1 gallon of gasoline. If gasoline costs 63 cents per gallon, what will be the cost of the gasoline required to drive 100 miles?

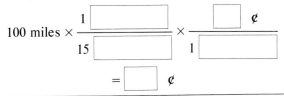

= □ ¢

Completed Problems

(a) If 12 pounds of bananas cost 63 cents, how much will 20 pounds cost?

$$\overset{5}{\cancel{20}}\,\cancel{lb} \times \frac{\overset{21}{\cancel{63}}\,¢}{\underset{1}{\underset{\cancel{A}}{\cancel{12}\,\cancel{lb}}}} = \boxed{105}\ ¢$$

(b) If 15 percent of a number is 63, what is 25 percent of the number?

$$\overset{5}{\cancel{25}\%} \times \frac{\overset{21}{\cancel{63}}}{\underset{1}{\underset{\cancel{5}}{\cancel{15}\,\%}}} = \boxed{105}$$

(c) A salesperson sells an average of 63 pairs of shoes every 6 days. If the shoes sell for $20 per pair, what is the expected total dollar value of sales over a 100-day period?

$$\overset{50}{\cancel{100}\,\text{days}} \times \frac{\overset{21}{\cancel{63}}\,\cancel{\text{pairs}}}{\underset{1}{\underset{2}{\cancel{6}\,\cancel{\text{days}}}}} \times \frac{\$\,\boxed{20}}{1\,\cancel{\text{pair}}} = \$\,\boxed{21{,}000}$$

(d) If there are 6 ounces of meat in 15 ounces of spaghetti, and meat costs 3 cents per ounce, what is the cost of the meat in 100 ounces of spaghetti?

$$\overset{20}{\cancel{100}\,\cancel{\text{oz spaghetti}}} \times \frac{\boxed{6}\,\cancel{\text{oz meat}}}{\underset{1}{\underset{\cancel{5}}{\cancel{15}\,\cancel{\text{oz spaghetti}}}}} \times \frac{\overset{1}{\cancel{3}}\,¢}{\boxed{1}\,\cancel{\text{oz meat}}} = \boxed{120}\ ¢$$

(e) A motorist can drive his car 15 miles on 1 gallon of gasoline. If gasoline costs 63 cents per gallon, what will be the cost of the gasoline required to drive 100 miles?

$$\overset{20}{\cancel{100}\,\cancel{\text{miles}}} \times \frac{1\ \boxed{\text{gallon}}}{\underset{1}{\underset{\cancel{5}}{\cancel{15}\ \boxed{\text{miles}}}}} \times \frac{\overset{21}{\cancel{63}}\,¢}{1\ \boxed{\text{gallon}}} = \boxed{420}\ ¢$$

Name: _____

Class: _____

Exercises for Sec. 11-1   Compute each product.

1. $8 \text{ lb} \times \dfrac{15¢}{\text{lb}} =$

2. $6 \text{ days} \times \dfrac{\$90}{4 \text{ days}} =$

3. $8 \text{ gal} \times \dfrac{\$6}{4 \text{ gal}} =$

4. $8 \text{ in} \times \dfrac{1 \text{ ft}}{12 \text{ in}} =$

5. $5 \text{ lb} \times \dfrac{16 \text{ oz}}{1 \text{ lb}} =$

6. $35 \times \dfrac{12\%}{27} =$

7. $63¢ \times \dfrac{\$1}{100¢} =$

8. $2 \text{ hr} \times \dfrac{50 \text{ gal}}{\text{hr}} \times \dfrac{\$3}{\text{gal}} =$

9. $5 \text{ refrigerators} \times \dfrac{12 \text{ men}}{15 \text{ refrigerators}} \times \dfrac{\$40}{\text{man}} =$

10. $42 \text{ in} \times \dfrac{1 \text{ ft}}{12 \text{ in}} \times \dfrac{1 \text{ yd}}{3 \text{ ft}} =$

11. $5 \text{ peaches} \times \dfrac{1 \text{ lb}}{3 \text{ peaches}} \times \dfrac{24¢}{\text{lb}} =$

12. $600 \text{ lb alloy} \times \dfrac{250 \text{ lb iron}}{700 \text{ lb alloy}} \times \dfrac{2{,}100 \text{ lb ore}}{500 \text{ lb iron}} =$

Work the problems using conversion factors.

13. 36 ounces is equal to how many pounds? (16 oz = 1 lb)

14. 3 pounds is equal to how many ounces? (16 oz = 1 lb)

15. 5 feet is equal to how many inches? (12 in = 1 ft)

16. 30 inches is equal to how many feet? (12 in = 1 ft)

17. If 15 pounds of sugar cost $1.65, how much will 9 pounds cost?

18. If 6 pounds of almonds cost 40¢, how much will 15 pounds cost?

19. If there are 5 ounces of meat in a 12-ounce can of ravioli, how many ounces of meat are there in an 18-ounce can?

20. If there are 6 grains of morphine in one pill, how many grains of morphine are there in ¼ pill?

21. If 4 men can lay 54 feet of sidewalk in 1 day, how many feet should 6 men be able to lay in 1 day?

22. A motorist finds her car can be driven 15 miles on 1 gallon of gasoline. How far can the car be driven on 6 gallons?

23. A motorist found that he used 15 gallons of gasoline to drive his car 275 miles. How much gasoline would be required to drive 924 miles?

24. For each share of stock, a company paid a dividend of $3.50 every 90 days. What would be the total dividend paid for one share after 540 days?

25. If a motorist drives 50 miles per hour, how long will it take her to drive 480 miles?

26. If a batter averages 60 hits for every 200 times at bat, how many hits would he be expected to get in 390 times at bat?

27. If a pitcher yields 20 runs in every 90 innings he pitches, how many runs would he be expected to yield in 180 innings?

28. A metal bar was found to expand at a rate of .0006 inch per 9-degree rise in temperature. How much would the bar expand if the temperature increased 48 degrees?

29. An investor with 60 shares of stock received a dividend check of $14.40. What dividend check should an investor with 100 shares receive?

30. There are 320 pounds of pure copper in 2,000 pounds of ore.
    (a) How much copper is there in 600 pounds of the ore?
    (b) How much ore would be required to produce 420 pounds of copper?

31. 12 percent of a number is 27.
    (a) What is 20 percent of the number?
    (b) 63 is what percent of the number?
    (c) What is 100 percent of the number?

32. If peaches are being sold for 22 cents per pound, and 7 peaches weigh 2 pounds, what would be the cost of 20 peaches?

33. A plane uses 50 gallons of fuel per hour. Fuel costs $1.20 per gallon. What will be the fuel cost if the plane is flown for 4 hours?

34. If water flows from a pipe at a rate of 80 gallons every 3 minutes, how much would flow out in 45 seconds? (60 sec = 1 min)

35. If 3 heads of lettuce sell for 80¢ and each head weighs 24 ounces, what is the price per ounce?

36. A motorist finds that he can drive his car 20 miles on 1 gallon of gasoline. If gasoline costs $.68 per gallon, what will be the fuel cost when the car is driven 150 miles?

37. In a given day, a company finds that 12 workers can assemble 56 air conditioners. If each worker is paid $40, what would be the labor cost to assemble 21 air conditioners?

38. A production line turns out refrigerators at a rate of 90 refrigerators per 8-hour day. The cost of producing one refrigerator is $270. What is the cost for 100 hours of production?

39. If there are 135 pounds of copper in 700 pounds of ore, and 50 pounds of copper in 90 pounds of brass:
    (a) How much ore is required to produce 729 pounds of brass?
    (b) How much brass would be produced from 5,600 pounds of ore?

40. If a box of apples costs $5.40 and apples are selling for $.12 per pound, and there are 5 apples per pound, how many apples are in one box?

Section 11-2 The Metric System

The metric system may be viewed as a base-10 system of measurement. Looking at the length and weight tables below, can you see why?

Length	*Weight*
1 kilometer = 1,000 meters	1 kilogram = 1,000 grams
1 hectometer = 100 meters	1 hectogram = 100 grams
1 dekameter = 10 meters	1 dekagram = 10 grams
1 meter (like the unit 1)	1 gram (like the unit 1)
1 decimeter = $\frac{1}{10}$ meter	1 decigram = $\frac{1}{10}$ gram
1 centimeter = $\frac{1}{100}$ meter	1 centigram = $\frac{1}{100}$ gram
1 millimeter = $\frac{1}{1,000}$ meter	1 milligram = $\frac{1}{1,000}$ gram

(*Note:* When each length or weight unit is multiplied by 10, the result is the unit of measure directly above.)

Some equivalences between metric measures and our English measures are given below.

  1 kilometer is approximately $\frac{6}{10}$ mile.

  1 meter is approximately 39 inches (1 yard + 3 inches).

  1 centimeter is approximately $\frac{2}{5}$ inch.   1 centimeter

  1 kilogram is approximately 2.2 pounds.

  450 grams are approximately 1 pound.

The length and weight equivalences that will be used in this unit are the common ones listed below. **Memorize them.**

Equivalences	Conversion Factors	
LENGTH		
1 kilometer (km) = 1,000 meters (m)	$\frac{1 \text{ km}}{1,000 \text{ m}}$ or	$\frac{1,000 \text{ m}}{1 \text{ km}}$
1 meter = 100 centimeters (cm)	$\frac{1 \text{ m}}{100 \text{ cm}}$ or	$\frac{100 \text{ cm}}{1 \text{ m}}$
1 meter = 1,000 millimeters (mm)	$\frac{1 \text{ m}}{1,000 \text{ mm}}$ or	$\frac{1,000 \text{ mm}}{1 \text{ m}}$
1 centimeter = 10 mm	$\frac{1 \text{ cm}}{10 \text{ mm}}$ or	$\frac{10 \text{ mm}}{1 \text{ cm}}$
WEIGHT		
1 kilogram (kg) = 1,000 grams (g)	$\frac{1 \text{ kg}}{1,000 \text{ g}}$ or	$\frac{1,000 \text{ g}}{1 \text{ kg}}$
1 gram = 1,000 milligrams (mg)	$\frac{1 \text{ g}}{1,000 \text{ mg}}$ or	$\frac{1,000 \text{ mg}}{1 \text{ g}}$

378 Denominate Numbers

Examples of Conversion

Length

(a) Convert 2,523 millimeters to meters.

Solution:
$$? \text{ m} = 2{,}523 \text{ mm} = 2{,}523 \text{ mm} \times 1$$
$$= 2{,}523 \text{ mm} \times \frac{1 \text{ m}}{1{,}000 \text{ mm}}$$
$$= \frac{2{,}523 \text{ m}}{1{,}000} = 2.523 \text{ m}$$

Length

(b) Convert .629 kilometer to centimeters.

Strategy: Find the equivalences that provide a "route" from kilometers to centimeters.

Equivalences:	Route:
1 km = 1,000 m	kilometers to meters
1 m = 100 cm	meters to centimeters

Solution:

route → km to m, m to cm

$$.629 \text{ km} \times \frac{\overbrace{1{,}000 \text{ m}}^{\text{km to m}}}{1 \text{ km}} \times \frac{\overbrace{100 \text{ cm}}^{\text{m to cm}}}{1 \text{ m}} = (.629)(100{,}000) \text{ cm} = 62{,}900 \text{ cm}$$

Weight

(c) Convert .278 kilogram to grams.

Solution:
$$.278 \text{ kg} \times \frac{1{,}000 \text{ g}}{1 \text{ kg}} = (.278)(1{,}000) \text{ g} = 278 \text{ g}$$

Weight

(d) Convert 1,253 milligrams to kilograms.

Strategy: Find a route from milligrams to kilograms.

Equivalences:	Route:
1,000 mg = 1 g	milligrams to grams
1,000 g = 1 kg	grams to kilograms

Solution:

route → mg to g, g to kg

$$1{,}253 \text{ mg} \times \frac{\overbrace{1 \text{ g}}^{\text{mg to g}}}{1{,}000 \text{ mg}} \times \frac{\overbrace{1 \text{ kg}}^{\text{g to kg}}}{1{,}000 \text{ g}} = \frac{1{,}253 \text{ kg}}{1{,}000{,}000} = .001253 \text{ kg}$$

Notice how easy it is to convert units in the metric system. Each conversion involves multiplication or division by a power of 10. The result of this multiplication or division is a shift of the decimal point to the right or to the left.

Area The **area of a rectangle** is defined to be the product of its length and width.

$$\text{Area} = \text{length} \times \text{width}$$
$$A = lw$$

A square is a rectangle with equal sides.

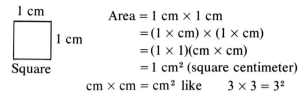

$$\begin{aligned}\text{Area} &= 1 \text{ cm} \times 1 \text{ cm} \\ &= (1 \times \text{cm}) \times (1 \times \text{cm}) \\ &= (1 \times 1)(\text{cm} \times \text{cm}) \\ &= 1 \text{ cm}^2 \text{ (square centimeter)}\end{aligned}$$

$$\text{cm} \times \text{cm} = \text{cm}^2 \text{ like } \quad 3 \times 3 = 3^2$$

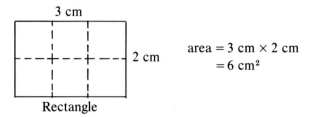

$$\begin{aligned}\text{area} &= 3 \text{ cm} \times 2 \text{ cm} \\ &= 6 \text{ cm}^2\end{aligned}$$

Examples

Area (a) Find the area of a rectangular field with length = 60 meters and width = 50 meters.

Solution:
$$\begin{aligned}A &= lw \\ &= 60 \text{ m} \times 50 \text{ m} \\ &= 3{,}000 \text{ m}^2 \quad \text{m} \times \text{m} = \text{m}^2\end{aligned}$$

Area (b) A rectangle has length = 80 centimeters and width = 30 centimeters. Find the area in square meters.

Solution:
$$A = 80 \text{ cm} \times 30 \text{ cm}$$

$$= 80 \cancel{\text{ cm}} \times 30 \cancel{\text{ cm}} \times \overbrace{\frac{1 \text{ m}}{100 \cancel{\text{ cm}}} \times \frac{1 \text{ m}}{100 \cancel{\text{ cm}}}}^{\text{cm}^2 \text{ to } \text{m}^2}$$

$$= \frac{2{,}400 \text{ m}^2}{10{,}000} = .24 \text{ m}^2$$

Whereas only one conversion factor is sufficient to convert centimeters to meters, two are needed to convert square centimeters to square meters. One could create conversion factors for converting squared units. For example,

$$1 \text{ m}^2 = 1 \cancel{\text{ m}} \times 1 \cancel{\text{ m}} \times \frac{100 \text{ cm}}{1 \cancel{\text{ m}}} \times \frac{100 \text{ cm}}{1 \cancel{\text{ m}}}$$

$$= 10{,}000 \text{ cm}^2$$

Conversion factor:

$$\frac{10{,}000 \text{ cm}^2}{1 \text{ m}^2} \quad \text{or} \quad \frac{1 \text{ m}^2}{10{,}000 \text{ cm}^2}$$

However, in the metric system, the process used in Example b is easy enough that memorizing the conversion factors above would not be worthwhile.

Example

Area   Convert 1 square kilometer to square centimeters.

Solution:

$$1 \text{ km}^2 = 1 \text{ km} \times 1 \text{ km} \times \overbrace{\frac{1{,}000 \text{ m}}{1 \text{ km}} \times \frac{1{,}000 \text{ m}}{1 \text{ km}}}^{\text{km}^2 \text{ to m}^2} \times \overbrace{\frac{100 \text{ cm}}{1 \text{ m}} \times \frac{100 \text{ cm}}{1 \text{ m}}}^{\text{m}^2 \text{ to cm}^2}$$

$$= 10{,}000{,}000{,}000 \text{ cm}^2$$

Volume   The **volume of a rectangular prism** (box) is defined to be the product of its length, width, and height.

$$\text{Volume} = \text{length} \times \text{width} \times \text{height}$$
$$V = lwh$$

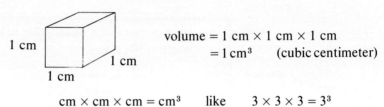

volume = 1 cm × 1 cm × 1 cm
= 1 cm³   (cubic centimeter)

cm × cm × cm = cm³   like   3 × 3 × 3 = 3³

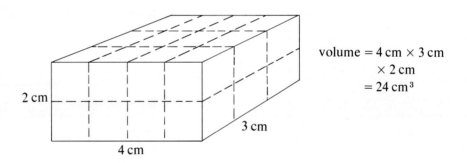

volume = 4 cm × 3 cm
× 2 cm
= 24 cm³

Special Measures of Volume

1 liter = 1,000 milliters (ml)
= 1,000 cm³
1 ml = 1 cm³

Examples

Volume

(a) A packing crate has length = 5 meters, width = 2 meters, and height = 3 meters. Find the volume in
  1. cubic meters
  2. cubic centimeters

Solution:

1. $V = 5\text{ m} \times 2\text{ m} \times 3\text{ m} = 30\text{ m}^3$
2. $V = 5\text{ m} \times 2\text{ m} \times 3\text{ m}$

$\qquad\qquad\overbrace{\phantom{m^3 \text{ to } cm^3}}^{m^3 \text{ to } cm^3}$

$= 5\,\cancel{m} \times 2\,\cancel{m} \times 3\,\cancel{m} \times \dfrac{100\text{ cm}}{1\,\cancel{m}} \times \dfrac{100\text{ cm}}{1\,\cancel{m}} \times \dfrac{100\text{ cm}}{1\,\cancel{m}}$

$= (30)(1{,}000{,}000)\text{ cm}^3$

$= 30{,}000{,}000\text{ cm}^3$

Volume

(b) Convert 1 liter to cubic meters.

Solution:

1 liter = 1,000 cm³

$\qquad\qquad\overbrace{\phantom{cm^3 \text{ to } m^3}}^{cm^3 \text{ to } m^3}$

$= 1{,}000 \times \cancel{cm} \times \cancel{cm} \times \cancel{cm} \times \dfrac{1\text{ m}}{100\,\cancel{cm}} \times \dfrac{1\text{ m}}{100\,\cancel{cm}} \times \dfrac{1\text{ m}}{100\,\cancel{cm}}$

$= \dfrac{1{,}000\text{ m}^3}{1{,}000{,}000}$

$= \dfrac{1}{1{,}000}\text{ m}^3 \quad$ or $\quad .001\text{ m}^3$

Addition and Subtraction with Mixed Units

Examples

(a) *Short form:*

```
   5 m    52 cm    6 mm
   7 m    23 cm    7 mm
+  2 m    39 cm    8 mm
  ────   ─────    ─────
  14 m   114 cm   21 mm
```

= 15 m   16 cm   1 mm

Expanded form:

```
   5 m   +    52 cm    +    6 mm
   7 m   +    23 cm    +    7 mm
+  2 m   +    39 cm    +    8 mm
  ────       ──────         ─────
  14 m   +   114 cm    +   21 mm
```

= 14 m + (1 m + 14 cm) + (2 cm + 1 mm)

= 15 m   + 16 cm   + 1 mm

The arrows indicate units that are being carried to the next column. Recall that 10 mm = 1 cm and 100 cm = 1 m.

(b)   7 kg   842 g   763 mg   *recall:* 1,000 mg = 1 g
      6 kg   546 g   397 mg   1,000 g = 1 kg
    + 8 kg   753 g   579 mg
     ─────────────────────────────
     21 kg   2,141 g   1,739 mg

   = 23 kg   142 g   739 mg

(c)   5 m   3̶2̶ cm   5̶ mm   borrow 1 cm = 10 mm
            31   15
     −3 m   27 cm   8 mm
     ─────────────────────────
      2 m   4 cm   7 mm

(d)   7̶ kg   3̶2̶5̶ g   4̶3̶2̶ mg   borrow 1 g = 1,000 mg
       6   1,324   1,432
     −2 kg   752 g   597 mg   and   1 kg = 1,000 g
     ─────────────────────────
      4 kg   572 g   835 mg

Sec. 11-2 The Metric System

Complete the problems below. Completed problems appear on page 384.

Sample Set

(a) Convert 623 millimeters to centimeters.

$$.623 \text{ mm} \times \frac{\boxed{} \text{ cm}}{\boxed{} \text{ mm}} = \boxed{} \text{ cm}$$

(b) Convert .592 kilogram to milligrams.

$$.592 \text{ kg} \times \frac{1{,}000 \boxed{}}{1 \boxed{}} \times \frac{\boxed{} \text{ mg}}{\boxed{} \text{ g}} = \boxed{} \text{ mg}$$

(c) A rectangle has length = 6 meters and width = 3 meters. Find the area in square centimeters.

$$A = 6 \text{ m} \times 3 \text{ m} \times \frac{\boxed{} \text{ cm}}{1 \boxed{}} \times \frac{\boxed{} \text{ cm}}{1 \boxed{}} = \boxed{} \text{ cm}^2$$

(d) Convert 1 cubic meter to liters.

$$1 \text{ m}^3 = 1 \text{ m} \times 1 \text{ m} \times 1 \text{ m} \times \frac{\boxed{}}{\boxed{}} \times \frac{\boxed{}}{\boxed{}}$$

$$\times \frac{\boxed{}}{\boxed{}} \times \frac{1 \text{ liter}}{\boxed{} \text{ cm}^3} = \boxed{} \text{ liters}$$

(e)

```
   7 kg    856 g    592 mg
   5 kg    967 g    776 mg
  +9 kg    523 g    892 mg
  ─────────────────────────
  ▢ kg   ▢ g     ▢ mg  =  ▢ kg  ▢ g  ▢ mg
```

(f)

```
   5 m   38 cm   5 mm
  -2 m   93 cm   8 mm
  ─────────────────────
  ▢ m   ▢ cm   ▢ mm
```

Completed Problems

(a) Convert 623 millimeters to centimeters.

$$.623 \text{ mm} \times \frac{1 \text{ cm}}{10 \text{ mm}} = \boxed{63.3} \text{ cm}$$

(b) Convert .592 kilogram to milligrams.

$$.592 \text{ kg} \times \frac{1{,}000 \text{ g}}{1 \text{ kg}} \times \frac{1000 \text{ mg}}{1 \text{ g}} = \boxed{592{,}000} \text{ mg}$$

(c) A rectangle has length = 6 meters and width = 3 meters. Find the area in square centimeters.

$$A = 6 \text{ m} \times 3 \text{ m} \times \frac{100 \text{ cm}}{1 \text{ m}} \times \frac{100 \text{ cm}}{1 \text{ m}} = \boxed{180{,}000} \text{ cm}^2$$

(d) Convert 1 cubic meter to liters.

$$1 \text{ m}^3 = 1 \text{ m} \times 1 \text{ m} \times 1 \text{ m} \times \frac{\overset{10}{\cancel{100 \text{ cm}}}}{1 \cancel{\text{ m}}} \times \frac{100 \text{ cm}}{1 \cancel{\text{ m}}}$$

$$\times \frac{\cancel{100 \text{ cm}}}{1 \cancel{\text{ m}}} \times \frac{1 \text{ liter}}{\underset{1}{\cancel{\underset{10}{\cancel{1000}}}} \text{ cm}^3} = \boxed{1{,}000} \text{ liters}$$

(e)
```
     7 kg    856 g    592 mg
     5 kg    967 g    776 mg
    +9 kg    523 g    892 mg
```
$\boxed{21}$ kg $\boxed{2346}$ g $\boxed{2260}$ mg = $\boxed{23}$ kg $\boxed{348}$ g $\boxed{260}$ mg

(f)
```
     5 m     38 cm    5 mm
    -2 m     93 cm    8 mm
```
$\boxed{2}$ m $\boxed{44}$ cm $\boxed{7}$ mm

Name: _____

Class: _____

Exercises for Sec. 11-2 Length

1. Convert .59 centimeter to millimeters.

2. Convert .623 meter to centimeters.

3. Convert 1,275 centimeters to meters.

4. Convert 837 millimeters to centimeters.

5. Convert 4,379 meters to kilometers.

6. Convert 5.293 kilometers to meters.

7. Convert .734 meter to millimeters.

8. Convert 8,372 centimeters to kilometers.

9. Convert 5,274,000 millimeters to kilometers.

10. Convert .00783 kilometer to meters.

Weight

11. Convert .053 gram to milligrams.

12. Convert .893 kilogram to grams.

13. Convert 5,325 grams to kilograms.

14. Convert 549 milligrams to grams.

15. Convert 458 milligrams to grams.

16. Convert .9546 gram to milligrams.

17. Convert .006512 kilogram to milligrams.

18. Convert .07635 kilogram to milligrams.

19. Convert 57,309,128 milligrams to kilograms.

20. Convert 3,417,390 milligrams to kilograms.

Area

21. Convert .728 square meter to square centimeters.

22. Convert 5,285 square millimeters to square centimeters.

23. Convert 5,852,931 square millimeters to square meters.

24. Convert .03841 square meter to square millimeters.

25. Convert 82,342,890 square meters to square kilometers.

26. Convert .00018241 square kilometer to square meters.

27. A rectangle has length = 3 meters and width = 2 meters. Find the area in square centimeters.

28. A rectangle has length = 60 centimeters and width = 20 centimeters. Find the area in square millimeters.

29. A rectangle has length = 5 millimeters and width = 4 millimeters. Find the area in square centimeters.

30. A rectangle has length = 50 centimeters and width = 30 centimeters. Find the area in square meters.

Volume

31. Convert 5 cubic kilometers to cubic meters.

32. Convert .008 cubic meter to cubic centimeters.

33. Convert 5,000 cubic millimeters to cubic centimeters.

34. Convert 8,000,000 cubic centimeters to cubic meters.

35. A prism has length = 3 meters, width = 2 meters, and height = 5 meters. Find the volume in cubic centimeters.

36. A rectangle has length = 5 centimeters, width = 2 centimeters, and height = 4 centimeters. Find the volume in cubic millimeters.

37. Convert 5,000 cubic millimeters to liters.

38. Convert 6 liters to cubic millimeters.

39. Convert 6,000 liters to cubic meters.

40. Convert .008 cubic meter to liters.

Addition and Subtraction

41.   5 m 72 cm 8 mm
    +2 m 53 cm 9 mm

42.   8 kg 732 g 591 mg
    +3 kg 547 g 823 mg

43.   7 kg 537 g 428 mg
   12 kg 652 g 795 mg
  + 8 kg 978 g 580 mg

44.   3 m 92 cm 5 mm
   6 m 62 cm 7 mm
  +4 m 39 cm 6 mm

45.   5 km   727 m   375 mm   46.   4 km   873 m   431 mm
          2 km   538 m   932 mm   9 km   460 m   698 mm
     +6 km   279 m   745 mm   +3 km   591 m   746 mm

47.   18 kg   347 g   896 mg   48.   7 m   56 cm   7 mm
   −11 kg   846 g   932 mg   −3 m   84 cm   6 mm

49.   9 m   84 cm   36 mm   50.   12 km   305 m   437 mm
   −8 m   94 cm   75 mm   − 8 km   732 m   863 mm

Section 11-3 The English System

The English system of measures is the system presently used in the United States. Equivalent measures that will be used in this unit are listed below:

English System

LENGTH
1 mile = 5,280 feet
1 mile = 1,760 yards
1 yard = 3 feet
1 foot = 12 inches

VOLUME
1 gallon = 231 cubic inches (exact)
1 gallon = 4 quarts
1 quart = 2 pints

WEIGHT
2,000 pounds = 1 ton
1 pound = 16 ounces

Examples of Conversion

Length

(a) Convert $\frac{1}{2}$ mile to feet.

Solution:
$$\frac{1}{2} \text{ mi} = \frac{1}{2} \text{ mi} \times 1$$
$$= \frac{1}{2} \cancel{\text{mi}} \times \frac{5{,}280 \text{ ft}}{\cancel{\text{mi}}}$$
$$= \frac{\overset{2{,}640}{\cancel{5{,}280}} \text{ ft}}{\underset{1}{\cancel{2}}} = 2{,}640 \text{ ft}$$

Length

(b) Convert 3,240 yards to miles.

Solution:
$$\cancel{3{,}240} \text{ yd} \times \frac{1 \text{ mi}}{\cancel{1{,}760} \text{ yd}} = \frac{81}{44} \text{ mi} = 1\frac{37}{44} \text{ mi}$$

Weight

(c) Convert 5,200 pounds to tons.

Solution:
$$\cancel{5{,}200} \text{ lb} \times \frac{1 \text{ ton}}{\cancel{2{,}000} \text{ lb}} = \frac{13}{5} \text{ tons} = 2\frac{3}{5} \text{ tons}$$

Weight

(d) Convert $\frac{1}{3}$ pound to ounces.

394 Denominate Numbers

Solution:
$$\frac{1}{3} \text{ lb} \times \frac{16 \text{ oz}}{1 \text{ lb}} = \frac{16}{3} \text{ oz} = 5\frac{1}{3} \text{ oz}$$

Area

(e) Convert 1 square yard to square feet.

Solution:
$$1 \text{ yd}^2 = 1 \text{ yd} \times 1 \text{ yd} \times \frac{3 \text{ ft}}{1 \text{ yd}} \times \frac{3 \text{ ft}}{1 \text{ yd}}$$
$$= 9 \text{ ft}^2$$

Area

(f) Convert 1 square foot to square inches.

Solution:
$$1 \text{ ft}^2 = 1 \text{ ft} \times 1 \text{ ft} \times \frac{12 \text{ in}}{1 \text{ ft}} \times \frac{12 \text{ in}}{1 \text{ ft}}$$
$$= 144 \text{ in}^2$$

It will be convenient to add the following equivalences to your list:

$$1 \text{ yd}^2 = 9 \text{ ft}^2$$
$$1 \text{ ft}^2 = 144 \text{ in}^2$$

Examples

Volume

(a) Convert 1 pint to cubic inches.

Strategy: Find a route from pints to cubic inches.

Equivalences:	*Route:*
2 pt = 1 qt	pints to quarts
4 qt = 1 gal	quarts to gallons
1 gal = 231 in³	gallons to cubic inches

Solution:

route → pt to qt, qt to gal, gal to in³

$$1 \text{ pt} \times \frac{1 \text{ qt}}{2 \text{ pt}} \times \frac{1 \text{ gal}}{4 \text{ qt}} \times \frac{231 \text{ in}^3}{1 \text{ gal}} = \frac{231 \text{ in}^3}{8} = 28\frac{7}{8} \text{ in}^3$$

Volume

(b) Convert 1 cubic foot to cubic inches.

Solution:
$$1 \text{ ft}^3 = 1 \text{ ft} \times 1 \text{ ft} \times 1 \text{ ft} \times \frac{12 \text{ in}}{1 \text{ ft}} \times \frac{12 \text{ in}}{1 \text{ ft}} \times \frac{12 \text{ in}}{1 \text{ ft}}$$
$$= 1{,}728 \text{ in}^3$$

Add the following equivalences to your list:

$$1 \text{ yd}^3 = 27 \text{ ft}^3$$
$$1 \text{ ft}^3 = 1{,}728 \text{ in}^3$$

Example

Volume

A rectangular tank has length = 6 yards, width = 3 yards, and height = 2 yards. How many gallons will the tank hold?

Strategy:

Equivalences:	*Route*
1 yd³ = 27 ft³	cubic yards to cubic feet
1 ft³ = 1,728 in³	cubic feet to cubic inches
231 in³ = 1 gal	cubic inches to gallons

Solution:

$$V = 6 \text{ yd} \times 3 \text{ yd} \times 2 \text{ yd}$$

$$= \underbrace{36 \text{ yd}^3} \times \frac{27 \text{ ft}^3}{1 \text{ yd}^3} \times \frac{1{,}728 \text{ in}^3}{1 \text{ ft}^3} \times \frac{1 \text{ gal}}{231 \text{ in}^3}$$

$$= \frac{(36)(27)(1{,}728)}{231} \text{ gal}$$

$$= \frac{1{,}679{,}616}{231} \text{ gal}$$

$$= 7{,}271.06 \text{ gal} \quad \text{to the nearest 100th}$$

The difficulty of the computation in the last example demonstrates the advantage of using the metric system. For example, if the tank has length = 6 meters, width = 3 meters, and height = 2 meters, the volume in liters is easily found.

$$V = 6 \text{ m} \times 3 \text{ m} \times 2 \text{ m} \times \frac{100 \text{ cm}}{1 \text{ m}} \times \frac{100 \text{ cm}}{1 \text{ m}} \times \frac{100 \text{ cm}}{1 \text{ m}} \times \frac{1 \text{ liter}}{1{,}000 \text{ cm}^3}$$

$$= \frac{(36)(100)(\cancel{100}^{10})(\cancel{100}^{1}) \text{ liter}}{\cancel{1{,}000}_{\cancel{10}_{1}}}$$

$$= 36{,}000 \text{ liters}$$

You will find it convenient to use a hand calculator when working most of the problems in this section.

Addition and Subtraction with Mixed Units

Examples

(a) Add:

$$\begin{array}{rrr} 2 \text{ yd} & 2 \text{ ft} & 10 \text{ in} \\ 1 \text{ yd} & 1 \text{ ft} & 9 \text{ in} \\ +\ 5 \text{ yd} & 2 \text{ ft} & 8 \text{ in} \\ \hline 8 \text{ yd} & 5 \text{ ft} & 27 \text{ in} \end{array}$$

= 8 yd 7 ft 3 in   carry 2 ft = 24 in to the ft column
= 10 yd 1 ft 3 in   carry 2 yd = 6 ft to the yd column

(b) Subtract:

$$\begin{array}{rr} \overset{4}{\cancel{5}} \text{ lb} & \overset{27}{\cancel{11}} \text{ oz} \\ -\ 2 \text{ lb} & 15 \text{ oz} \\ \hline 2 \text{ lb} & 12 \text{ oz} \end{array}$$

borrow 16 oz = 1 lb from the lb column; then
16 oz + 11 oz = 27 oz

Complete the problems below. Completed problems appear on page 398.

Sample Set

(a) Convert 5 yards to inches.

$$5 \text{ yds} \times \frac{\boxed{} \text{ ft}}{\boxed{} \text{ yd}} \times \frac{\boxed{} \text{ in}}{\boxed{} \text{ ft}} = \boxed{} \text{ in}$$

(b) Convert 28,000 ounces to tons.

$$28{,}000 \text{ oz} \times \frac{1 \boxed{}}{16 \boxed{}} \times \frac{1 \boxed{}}{2{,}000 \boxed{}} = \boxed{} \text{ ton}$$

(c) Convert 10 gallons to cubic feet (nearest 10th).

$$10 \text{ gal} \times \frac{\boxed{}}{\boxed{}} \times \frac{\boxed{}}{\boxed{}} = \boxed{} \text{ ft}^3$$

(d) A rectangle has length = 10 yards and width = 2 yards. Find the area in square inches.

$$A = 10 \text{ yds} \times 2 \text{ yds} \times \frac{\boxed{} \text{ ft}^2}{\boxed{} \text{ yd}^2} \times \frac{\boxed{} \text{ in}^2}{\boxed{} \text{ ft}^2} = \boxed{} \text{ in}^2$$

(e) A prism has length = 12 inches, width = 6 inches, and height = 9 inches. Find the volume in cubic yards (as a fraction in lowest terms).

$$V = 12 \text{ in} \times 6 \text{ in} \times 9 \text{ in} \times \frac{\boxed{}}{\boxed{}} \times \frac{\boxed{}}{\boxed{}} = \boxed{} \text{ yd}^3$$

Completed Problems

(a) Convert 5 yards to inches.

$$5 \text{ yds} \times \frac{\boxed{3} \text{ ft}}{\boxed{1} \text{ yd}} \times \frac{\boxed{12} \text{ in}}{\boxed{1} \text{ ft}} = \boxed{180} \text{ in}$$

(b) Convert 28,000 ounces to tons.

$$\overset{7}{\underset{8}{\cancel{\underset{\cancel{14}}{\cancel{28,000}}}}} \text{ oz} \times \frac{1 \boxed{\text{lb.}}}{16 \boxed{\text{oz.}}} \times \frac{1 \boxed{\text{ton}}}{2{,}000 \boxed{\text{lb.}}} = \boxed{\tfrac{7}{8}} \text{ ton}$$

(c) Convert 10 gallons to cubic feet (nearest 10th).

$$10 \text{ gal} \times \frac{\boxed{231 \text{ in}^3}}{\boxed{1 \text{ gal.}}} \times \frac{\boxed{1 \text{ ft}^3}}{\boxed{1{,}728 \text{ in}^3}} = \boxed{1.34} \text{ ft}^3$$

(d) A rectangle has length = 10 yards and width = 2 yards. Find the area in square inches.

$$A = 10 \text{ yds} \times 2 \text{ yds} \times \frac{\boxed{9} \text{ ft}^2}{\boxed{1} \text{ yd}^2} \times \frac{\boxed{144} \text{ in}^2}{\boxed{1} \text{ ft}^2} = \boxed{25{,}920} \text{ in}^2$$

(e) A prism has length = 12 inches, width = 6 inches, and height = 9 inches. Find the volume in cubic yards (as a fraction in lowest terms).

$$V = \overset{1}{\underset{\cancel{72}}{\cancel{\underset{432}{\cancel{12}}}}} \text{ in} \times \overset{1}{\cancel{6}} \text{ in} \times \overset{1}{\cancel{9}} \text{ in} \times \frac{\boxed{1 \text{ ft}^3}}{\boxed{1728 \text{ in}^3}} \times \frac{\boxed{1 \text{ yd}^3}}{\boxed{27 \text{ ft}^3}} = \boxed{\tfrac{1}{72}} \text{ yd}^3$$

Name: _____

Class: _____

Exercises for Sec. 11-3 Length

1. Convert $\frac{5}{6}$ foot to inches.

2. Convert 16 inches to feet.

3. Convert 7,040 feet to miles.

4. Convert $\frac{2}{3}$ mile to feet.

5. Convert $\frac{7}{8}$ yard to inches.

6. Convert 54 inches to yards.

7. Convert $\frac{1}{180}$ mile to inches.

8. Convert 42,240 inches to miles.

Weight

9. Convert 36 ounces to pounds.

10. Convert $\frac{5}{12}$ pound to ounces.

11. Convert $\frac{7}{24}$ pound to ounces.

12. Convert 42 ounces to pounds.

13. Convert $\frac{3}{4}$ ton to pounds.

14. Convert 2,500 pounds to tons.

15. Convert 19,200 ounces to tons.

16. Convert $\frac{5}{8}$ ton to ounces.

Area

17. Convert 720 square inches to square feet.

18. Convert 216 square inches to square yards.

19. Convert $\frac{1}{54}$ square yard to square inches.

20. Convert $\frac{1}{54}$ square foot to square inches.

21. Convert 1 square mile to square feet.

22. Convert 1 square mile to square yards.

23. A rectangle has length = 3 yards and width = 2 yards. Find the area in square feet.

24. A rectangle has length = 5 feet and width = 2 feet. Find the area in square inches.

Volume

25. Convert $\frac{1}{12}$ cubic yard to cubic feet.

26. Convert $\frac{1}{144}$ cubic foot to cubic inches.

27. Convert 14,580 cubic inches to cubic yards.

28. Convert $\frac{1}{144}$ cubic yards to cubic inches.

29. Convert $\frac{1}{12}$ gallon to pints.

30. Convert 38 pints to gallons.

31. Convert 72 gallons to cubic feet (nearest 10th).

32. Convert 540 gallons to cubic yards (nearest 10th).

Addition and Subtraction

33.   5 gal   3 qt   1 pt
      2 gal   2 qt   1 pt
   + 4 gal   1 qt   1 pt

34.   3 yd   2 ft   11 in
      2 yd   2 ft   3 in
   + 1 yd   2 ft   7 in

35.   8 yd   2 ft   10 in
      3 yd   1 ft   8 in
   + 7 yd   1 ft   9 in

36.   3 gal   2 qt   1 pt
      5 gal   2 qt
   + 3 gal   1 qt   1 pt

37.   5 lb   10 oz
   − 3 lb   15 oz

38.   12 tons   525 lb
   − 8 tons   725 lb

39.   7 yd   2 ft   5 in
   − 4 yd   2 ft   7 in

40.   4 yd   1 ft   7 in
   − 2 yd   2 ft   11 in

Section 11-4 English to Metric and Metric to English

Equivalent units that will be used in this section are listed below.

LENGTH
1 inch = 2.54 centimeters

AREA
1 square inch = 6.45 square centimeters approximate

WEIGHT
2.20 pounds = 1 kilogram approximate*

VOLUME
1 cubic inch = 16.39 cubic centimeters approximate

---

*The zero in the hundredths' place indicates the approximation has been rounded to the nearest 100th.

In this section, it will also be necessary to use the equivalences introduced in Secs. 11-2 and 11-3. A table of equivalent measures can be found at the end of the section (page 408). As new equivalences are found in the examples, you may find it convenient to add them to the list.

Examples

Length

(a) Convert 1 centimeter to inches.

Solution:

$$1 \text{ cm} \times \frac{1 \text{ in}}{2.54 \text{ cm}} = \frac{1 \text{ in}}{2.54} = .39 \text{ in} \quad \text{to the nearest 100th}$$

Length

(b) Convert 1 mile to kilometers.

Strategy: Find a route from miles to kilometers.

Equivalences:	Route:
1 mi = 5,280 ft	miles to feet
1 ft = 12 in	feet to inches
1 in = 2.54 cm	inches to centimeters
100 cm = 1 m	centimeters to meters
1,000 m = 1 km	meters to kilometers

Solution:

$$1 \text{ mi} \times \frac{5,280 \text{ ft}}{1 \text{ mi}} \times \frac{12 \text{ in}}{1 \text{ ft}} \times \frac{2.54 \text{ cm}}{1 \text{ in}} \times \frac{1 \text{ m}}{100 \text{ cm}} \times \frac{1 \text{ km}}{1,000 \text{ m}}$$

$$= \frac{(5,280)(12)(2.54)}{(100)(1,000)} \text{ km}$$

$$= 1.61 \text{ km} \quad \text{to the nearest 100th}$$

Weight

(c) Convert 453.59 grams to pounds.

Equivalences:	Route:
1,000 g = 1 kg	grams to kilograms
1 kg = 2.20 lb	kilograms to pounds

405

Solution:

$$453.59 \;\cancel{g} \times \frac{1 \;\cancel{kg}}{1{,}000 \;\cancel{g}} \times \frac{2.20 \text{ lb}}{1 \;\cancel{kg}} = \frac{(453.59)(2.20)}{1{,}000} \text{ lb}$$

$$= \frac{997.898}{1{,}000} \text{ lb}$$

$$= .997898 \text{ lb}$$

$$= 1.00 \text{ lb} \quad \text{to the nearest 100th}$$

Weight   (d) Convert 1 ounce to grams.

Equivalences:	Route:
16 oz = 1 lb	ounces to pounds
2.20 lb = 1 kg	pounds to kilograms
1 kg = 1,000 g	kilograms to grams

Solution:

$$1 \;\cancel{oz} \times \frac{1 \;\cancel{lb}}{16 \;\cancel{oz}} \times \frac{1 \;\cancel{kg}}{2.20 \;\cancel{lb}} \times \frac{1{,}000 \text{ g}}{1 \;\cancel{kg}} = \frac{1{,}000 \text{ g}}{(16)(2.20)} = 28.4 \text{ g}$$

Area   (e) Convert 1 square inch to square centimeters.

Solution:

$$1 \text{ in}^2 = 1 \;\cancel{in} \times 1 \;\cancel{in} \times \frac{2.54 \text{ cm}}{1 \;\cancel{in}} \times \frac{2.54 \text{ cm}}{1 \;\cancel{in}} = 6.4516 \text{ cm}^2$$

$$= 6.45 \text{ cm}^2 \quad \text{approximate}$$

Area   (f) Convert 1 square meter to square feet.

Solution:

$$1 \text{ m}^2 = 1 \;\cancel{m} \times 1 \;\cancel{m} \times \overbrace{\frac{100 \;\cancel{cm}}{1 \;\cancel{m}} \times \frac{100 \;\cancel{cm}}{1 \;\cancel{m}}}^{m^2 \text{ to } cm^2} \times \overbrace{\frac{1 \;\cancel{in^2}}{6.45 \;\cancel{cm^2}}}^{cm^2 \text{ to } in^2} \times \overbrace{\frac{1 \text{ ft}^2}{144 \;\cancel{in^2}}}^{in^2 \text{ to } ft^2}$$

$$= \frac{(100)(100) \text{ ft}^2}{(6.45)(144)}$$

$$= 10.77 \text{ ft}^2 \quad \text{nearest 100th}$$

Volume   (g) Convert 28,322 cubic centimeters to cubic feet.

Solution:

$$28{,}322 \;\cancel{cm^3} \times \frac{1 \;\cancel{in^3}}{16.39 \;\cancel{cm^3}} \times \frac{1 \text{ ft}^3}{1{,}728 \;\cancel{in^3}} = \frac{28{,}322 \text{ ft}^3}{(16.39)(1{,}728)}$$

$$= 1.00 \text{ ft}^3 \quad \text{nearest 100th}$$

Volume   (h) Convert 1 gallon to liters.

Solution:

$$1 \;\cancel{gal} \times \frac{231 \;\cancel{in^3}}{1 \;\cancel{gal}} \times \frac{16.39 \;\cancel{cm^3}}{1 \;\cancel{in^3}} \times \frac{1 \text{ liter}}{1{,}000 \;\cancel{cm^3}} = \frac{(231)(16.39) \text{ liters}}{1{,}000}$$

$$= 3.79 \text{ liters} \quad \text{nearest 100th}$$

Sec. 11-4 English to Metric and Metric to English

Complete the problems below. Completed problems appear on page 409.

Sample Set

(a) Convert 5 feet to centimeters.

$$5 \text{ ft} \times \frac{\boxed{} \text{ in}}{\boxed{} \text{ ft}} \times \frac{\boxed{} \text{ cm}}{\boxed{} \text{ in}} = \boxed{} \text{ cm}$$

(b) Convert 12 ounces to kilograms (nearest 100th).

$$12 \text{ oz} \times \frac{\boxed{}}{\boxed{}} \times \frac{1 \text{ kg}}{\boxed{} \text{ lb}} = \boxed{} \text{ kg}$$

(c) Convert 36 square centimeters to square feet (nearest 100th).

$$36 \text{ cm}^2 \times \frac{1 \text{ in}^2}{\boxed{}} \times \frac{1 \text{ ft}^2}{\boxed{}} = \boxed{} \text{ ft}^2$$

(d) Convert 108 cubic centimeters to cubic feet (nearest 1,000th).

$$108 \text{ cm}^3 \times \frac{\boxed{}}{\boxed{}} \times \frac{\boxed{}}{\boxed{}} = \boxed{} \text{ ft}^3$$

(e) Convert 1 pint to liters (nearest 100th).

$$1 \text{ pt} \times \frac{1 \text{ qt}}{\boxed{}} \times \frac{1 \text{ gal}}{\boxed{}} \times \frac{\boxed{}}{1 \text{ gal}}$$

$$\times \frac{\boxed{}}{1 \text{ in}^3} \times \frac{1 \text{ liter}}{\boxed{}} = \boxed{} \text{ liter}$$

Table of Equivalences

Metric System

LENGTH
1 kilometer (km) = 1,000 meters (m)
1 meter (m) = 100 centimeters (cm)
1 centimeter (cm) = 10 millimeters (mm)

WEIGHT
1 kilogram (kg) = 1,000 grams (g)
1 gram (g) = 1,000 milligrams (mg)

VOLUME
1 liter = 1,000 milliliters (ml)
1 milliliter (ml) = 1 cubic centimeter (cm^3)

English System

LENGTH
1 mile (mi) = 5,280 feet (ft)
1 mile = 1,760 yards (yd)
1 yard (yd) = 3 feet (ft)
1 foot (ft) = 12 inches (in)

WEIGHT
1 ton = 2,000 pounds (lb)
1 pound (lb) = 16 ounces (oz)

AREA
1 square yard (yd^2) = 9 square feet (ft^2)
1 square foot (ft^2) = 144 square inches (in^2)

VOLUME
1 gallon (gal) = 231 cubic inches (in^3)
1 gallon (gal) = 4 quarts (qt)
1 quart (qt) = 2 pints (pt)
1 cubic yard (yd^3) = 27 cubic feet (ft^3)
1 cubic foot (ft^3) = 1,728 cubic inches (in^3)

Equivalences between English and Metric

LENGTH
1 in = 2.54 cm

AREA
1 in^2 = 6.45 cm^2 approximate

WEIGHT
2.20 lb = 1 kg   approximate

VOLUME
1 in^3 = 16.39 cm^3 approximate

Completed Problems

(a) Convert 5 feet to centimeters.

$$5 \text{ ft} \times \frac{\boxed{12} \text{ in}}{\boxed{1} \text{ ft}} \times \frac{\boxed{2.54} \text{ cm}}{\boxed{1} \text{ in}} = \boxed{152.4} \text{ cm}$$

(b) Convert 12 ounces to kilograms (nearest 100th).

$$\overset{3}{\cancel{12 \text{ oz}}} \times \frac{\boxed{1 \text{ lb}}}{\underset{4}{\boxed{16 \text{ oz}}}} \times \frac{1 \text{ kg}}{\boxed{2.20} \text{ lb}} = \boxed{.34} \text{ kg}$$

(c) Convert 36 square centimeters to square feet (nearest 100th).

$$\overset{1}{\cancel{36 \text{ cm}^2}} \times \frac{1 \text{ in}^2}{\boxed{6.45 \text{ cm}^2}} \times \frac{1 \text{ ft}^2}{\underset{4}{\boxed{144 \text{ in}^2}}} = \boxed{.04} \text{ ft}^2$$

(d) Convert 108 cubic centimeters to cubic feet (nearest 1,000th).

$$\overset{1}{\cancel{108 \text{ cm}^3}} \times \frac{\boxed{1 \text{ in}^3}}{\boxed{16.39 \text{ cm}^3}} \times \frac{\boxed{1 \text{ ft}^3}}{\underset{16}{\boxed{1728 \text{ in}^3}}} = \boxed{.004} \text{ ft}^3$$

(e) Convert 1 pint to liters (nearest 100th).

$$1 \text{ pt} \times \frac{1 \text{ qt}}{\boxed{2 \text{ pt}}} \times \frac{1 \text{ gal}}{\boxed{4 \text{ qt}}} \times \frac{\boxed{231 \text{ in}^3}}{1 \text{ gal}}$$

$$\times \frac{\boxed{16.39 \text{ cm}^3}}{1 \text{ in}^3} \times \frac{1 \text{ liter}}{\boxed{1,000 \text{ cm}^3}} = \boxed{.47} \text{ liter}$$

Name: _____

Class: _____

Exercises for Sec. 11-4 Length

1. Convert 2 centimeters to inches (nearest 100th).

2. Convert 3 inches to centimeters.

3. Convert 5 inches to millimeters.

4. Convert 5 millimeters to inches (nearest 10th).

5. Convert 25 inches to meters.

6. Convert 5 meters to inches (nearest 10th).

7. Convert 12 meters to feet (nearest 10th).

8. Convert 10 yards to meters (nearest 10th).

9. Convert 100 miles to kilometers (nearest 10th).

10. Convert 1,056 kilometers to miles (nearest 10th).

Weight

11. Convert 6 kilograms to pounds.

12. Convert 4 pounds to kilograms (nearest 10th).

13. Convert 22 pounds to grams.

14. Convert 2,000 grams to pounds (nearest 10th).

15. Convert 12 ounces to kilograms (nearest 100th).

16. Convert ¼ kilogram to ounces.

17. Convert 5 grams to ounces.

18. Convert 32 ounces to grams (nearest whole unit).

19. Convert 1 ounce to milligrams (nearest whole unit).

20. Convert 2,000 milligrams to ounces.

Area

21. Convert 2 square inches to square centimeters.

22. Convert 5 square centimeters to square inches (nearest 100th).

23. Convert 1,000 square millimeters to square inches (nearest 10th).

24. Convert 1 square inch to square millimeters.

25. Convert 2,000 square inches to square meters.

26. Convert 3 square meters to square inches (nearest 10th).

27. Convert 1 square foot to square centimeters.

28. Convert 72 square centimeters to square feet (nearest 100th).

29. Convert 1 square yard to square meters (nearest 100th).

30. Convert 1 square meter to square yards (nearest 10th).

Volume

31. Convert 100 cubic centimeters to cubic inches (nearest 100th).

32. Convert 10 cubic inches to cubic centimeters.

33. Convert 1 cubic foot to cubic centimeters.

34. Convert 34,560 cubic centimeters to cubic feet (nearest 10th).

35. Convert 20,000 cubic millimeters to cubic inches (nearest 10th).

36. Convert 1,000 cubic inches to cubic millimeters.

37. Convert 1 quart to liters (nearest 10th).

38. Convert 1 liter to quarts (nearest 100th).

39. Convert 32 liters to cubic feet (nearest 10th).

40. Convert 1 cubic foot to liters (nearest 10th).

Section 11-5 Applications: Speed, Density, Concentration, and Cost per Unit

Speed is defined to the distance traveled per unit of time. Two examples are given below.

$$\frac{55 \text{ mi}}{1 \text{ hr}} = 55 \text{ mi/hr}$$

$$\frac{186{,}000 \text{ mi}}{1 \text{ sec}} = 186{,}000 \text{ mi/sec} = \text{speed of light}$$

Examples

Speed

(a) Convert 60 miles per hour to feet per hour.

Solution:

$$\frac{60 \text{ mi}}{1 \text{ hr}} \times \frac{5{,}280 \text{ ft}}{1 \text{ mi}} = \frac{(60)(5{,}280) \text{ ft}}{1 \text{ hr}} = 316{,}800 \text{ ft/hr}$$

Speed

(b) Convert 60 miles per hour to feet per second.

Solution:

In this problem, miles must be converted to feet, and hours must be converted to seconds. The effect of multiplying by the conversion factors is indicated above each.

$$\frac{60 \text{ mi}}{1 \text{ hr}} \times \overbrace{\frac{5{,}280 \text{ ft}}{1 \text{ mi}}}^{\text{to ft/hr}} \times \overbrace{\frac{1 \text{ hr}}{60 \text{ min}}}^{\text{to ft/min}} \times \overbrace{\frac{1 \text{ min}}{60 \text{ sec}}}^{\text{to ft/sec}} = \frac{\overset{1}{(60)}(5{,}280)^{88} \text{ ft}}{\underset{1}{(60)}\underset{1}{(60)} \text{ sec}} = 88 \text{ ft/sec}$$

Or, in a different order,

$$\frac{60 \text{ mi}}{1 \text{ hr}} \times \overbrace{\frac{1 \text{ hr}}{60 \text{ min}}}^{\text{to mi/min}} \times \overbrace{\frac{1 \text{ min}}{60 \text{ sec}}}^{\text{to mi/sec}} \times \overbrace{\frac{5{,}280 \text{ ft}}{1 \text{ mi}}}^{\text{to ft/sec}} = 88 \text{ ft/sec}$$

We can conclude that a car traveling 60 miles per hour travels 88 feet each second.

Speed

(c) Convert 1 inch per minute to centimeters per second.

Solution:

$$\frac{1 \text{ in}}{1 \text{ min}} \times \frac{1 \text{ min}}{60 \text{ sec}} \times \frac{2.54 \text{ cm}}{1 \text{ in}} = \frac{2.54 \text{ cm}}{60 \text{ sec}} = .04 \text{ cm/sec}$$

Speed

(d) If a car travels 50 miles per hour, how far will the car travel in 5 hours?

Solution:

$$5 \text{ hr} \times \frac{50 \text{ mi}}{1 \text{ hr}} = 250 \text{ mi}$$

418 Denominate Numbers

Speed

(e) If a plane travels 500 miles per hour, how many kilometers will it travel in
1. 2 hours?
2. 45 minutes?

Use 1 mile = 1.61 kilometers.

Solution:

1. $2 \, \cancel{hr} \times \dfrac{500 \, \cancel{mi}}{1 \, \cancel{hr}} \times \dfrac{1.61 \, km}{1 \, \cancel{mi}} = 1{,}610 \, km$

2. $\cancel{45}^{\,3} \, \cancel{min} \times \dfrac{1 \, \cancel{hr}}{\cancel{60}_{\,4} \, \cancel{min}} \times \dfrac{\cancel{500}^{\,125} \, \cancel{mi}}{1 \, \cancel{hr}} \times \dfrac{1.61 \, km}{1 \, \cancel{mi}}$

   $= 604 \, km$ nearest whole unit

Density The **density** (weight density) of a substance is defined to be its weight per unit volume.

Examples   Cast iron: $\dfrac{450 \, lb}{1 \, ft^3} = 450 \, lb/ft^3$

Gold: $\dfrac{19.3 \, g}{1 \, cm^3} = 19.3 \, g/cm^3$

Water: $\dfrac{1 \, g}{1 \, cm^3} = 1 \, g/cm^3$

Examples

Density (a) Convert 450 pounds per cubic foot to ounces per cubic inch.

Solution: $\dfrac{450 \, \cancel{lb}}{1 \, \cancel{ft^3}} \times \dfrac{16 \, oz}{1 \, \cancel{lb}} \times \dfrac{1 \, \cancel{ft^3}}{1{,}728 \, in^3} = \dfrac{(450)(16) \, oz}{1{,}728 \, in^3}$

$= 4.17 \, oz/in^3$

Density (b) Convert 1 gram per cubic centimeter to ounces per cubic inch.

Solution: $\dfrac{1 \, g}{1 \, \cancel{cm^3}} \times \dfrac{16.39 \, \cancel{cm^3}}{1 \, in^3} \times \dfrac{1 \, \cancel{kg}}{1{,}000 \, \cancel{g}} \times \dfrac{2.20 \, \cancel{lb}}{1 \, \cancel{kg}} \times \dfrac{16 \, oz}{1 \, \cancel{lb}} = \dfrac{(16.39)(2.20)(16) \, oz}{1{,}000 \, in^3}$

$= .577 \, oz/in^3$

Density (c) Convert 1 kilogram per cubic meter to grams per cubic centimeter.

Solution:

$\dfrac{1 \, \cancel{kg}}{\cancel{m^3}} \times \dfrac{1 \, \cancel{m}}{\cancel{100}_{\,1} \, cm} \times \dfrac{1 \, \cancel{m}}{\cancel{100}_{\,10} \, cm} \times \dfrac{1 \, \cancel{m}}{100 \, cm} \times \dfrac{\cancel{1{,}000}^{\,1} \, g}{1 \, \cancel{kg}} = \dfrac{1 \, g}{1{,}000 \, cm^3}$

$= .001 \, g/cm^3$

Sec. 11-5 Applications: Speed, Density, Concentration, and Cost per Unit

Density

(d) The volume of a crown was found to be 400 cubic centimeters. If the crown weighed 7,170 grams, could it be pure gold?

Solution: Reduce 7,170 g per 400 cm³ to grams per cubic centimeter.

$$\frac{7{,}170 \text{ g}}{400 \text{ cm}^3} = 17.925 \text{ g/cm}^3$$

Since the density of gold is 19.3 g/cm³, the crown is not pure gold.

Density

(e) Using the density 19.3 grams per cubic centimeter, find the weight of 1,000 cubic centimeters of gold in kilograms.

Strategy: Multiply 1,000 cubic centimeters by conversion factors that will cancel cubic centimeters and introduce kilograms.

Solution:

$$1{,}000 \text{ cm}^3 \times \frac{19.3 \text{ g}}{1 \text{ cm}^3} \times \frac{1 \text{ kg}}{1{,}000 \text{ g}} = 19.3 \text{ kg}$$

Density

(f) Using the density 19.3 grams per cubic centimeter, find the weight of 1 cubic inch of gold in ounces.

Solution:

$$1 \text{ in}^3 \times \frac{16.39 \text{ cm}^3}{1 \text{ in}^3} \times \frac{19.3 \text{ g}}{1 \text{ cm}^3} \times \frac{1 \text{ kg}}{1{,}000 \text{ g}} \times \frac{2.20 \text{ lb}}{1 \text{ kg}} \times \frac{16 \text{ oz}}{1 \text{ lb}}$$

$$= \frac{(16.39)(19.3)(2.20)(16) \text{ oz}}{1{,}000} = 11.13 \text{ oz}$$

Conclusion: The density of gold in the English system is 11.13 ounces per cubic inch.

Concentration

The **concentration** (weight concentration) of a substance in a liquid is defined to be the weight of that substance in a unit volume of the liquid. For example, if 1.5 grams of atropine sulfate were dissolved in 1 liter of pure water, the concentration of the atropine sulfate would be 1.5 grams per liter of solution.

Examples

Concentration

(a) Convert the concentration 1.5 grams per liter to milligrams per cubic centimeter.

Solution:

$$\frac{1.5 \text{ g}}{1 \text{ liter}} \times \frac{1 \text{ liter}}{1{,}000 \text{ cm}^3} \times \frac{1{,}000 \text{ mg}}{1 \text{ g}} = \frac{1.5 \text{ mg}}{1 \text{ cm}^3} = 1.5 \text{ mg/cm}^3$$

The beautiful result one would expect in the metric system.
Interpretation: If there are 1.5 grams in each liter of solution, then there are 1.5 milligrams in each cubic centimeter of that solution.

Concentration

(b) Convert 5 ounces per gallon into ounces per cubic inch.

Solution:

$$\frac{5 \text{ oz}}{1 \text{ gal}} \times \frac{1 \text{ gal}}{231 \text{ in}^3} = \frac{5 \text{ oz}}{231 \text{ in}^3} = .02 \text{ oz/in}^3$$

Concentration

(c) *(For nurses)* A doctor orders a dosage of .5 milligram of atropine sulfate to be given to a patient. The nurse finds a stock solution with the concentration 3 milligrams/2 cubic centimeters on the bottle. How many cubic centimeters of this solution should the nurse give the patient?

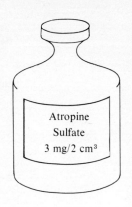

Strategy: Multiply .5 milligram by conversion factors that cancel milligram and introduce cubic centimeters.

$$.5 \text{ mg} \times \frac{2 \text{ cm}^3}{3 \text{ mg}} = \frac{1 \text{ cm}^3}{3} = .33 \text{ cm}^3 \quad \text{answer}$$

(*Note:* $.5 \times 2 = 1$)

Concentration

(d) In example c, if the doctor had ordered the dosage given in minims, how many minims should the nurse give? [1 minim (m̚) = 1 drop, and 15 minims = 1 cubic centimeter.]

Solution:

$$.5 \text{ mg} \times \frac{2 \text{ cm}^3}{\cancel{3} \text{ mg}} \times \frac{\overset{5}{\cancel{15}}\text{m̚}}{1 \text{ cm}^3} = (.5)(2)(5) \text{ m̚} = 5 \text{ m̚} \quad \text{answer}$$

Cost Rates

(a) If carpet costs $6 per square yard, find the cost in cents per square foot.

Solution:

$$\frac{\overset{2}{\cancel{\$6}}}{1 \text{ yd}^2} \times \frac{1 \text{ yd}^2}{\underset{3}{\cancel{9}} \text{ ft}^2} \times \frac{100\cancel{c}}{\cancel{\$1}} = \frac{200\cancel{c}}{3 \text{ ft}^2} = 66\frac{2}{3}\cancel{c}/\text{ft}^2$$

Cost Rates

(b) If gold costs $170 per ounce, find the cost in dollars per gram.

Solution:

$$\frac{\$170}{1 \text{ oz}} \times \frac{16 \text{ oz}}{1 \text{ lb}} \times \frac{2.20 \text{ lb}}{1 \text{ kg}} \times \frac{1 \text{ kg}}{1,000 \text{ g}} = \frac{(\$170)(16)(2.20)}{1,000 \text{ g}}$$

$$= \frac{\$5,984}{1,000 \text{ g}}$$

$$= \$5.98/\text{g}$$

Cost Rates

(c) If carpet costs $12 per square yard, find the cost of 100 square feet of carpet.

Strategy: Multiply 100 square feet by conversion factors that cancel square feet and introduce dollars.

Solution:

$$100 \text{ ft}^2 \times \frac{1 \text{ yd}^2}{\underset{9}{\cancel{27}} \text{ ft}^2} \times \frac{\overset{4}{\cancel{\$12}}}{1 \text{ yd}^2} = \frac{\$400}{9} = \$44.44$$

Cost Rates

(d) If milk costs 56 cents per quart, find the cost of 1 liter of milk. (Use 1 liter = 1.06 quarts.)

Sec. 11-5 Applications: Speed, Density, Concentration, and Cost per Unit

Solution:

$$1 \text{ liter} \times \frac{1.06 \text{ qt}}{1 \text{ liter}} \times \frac{56¢}{1 \text{ qt}} = 59¢ \quad \text{nearest cent}$$

Cost Rates

(e) Using the rate $5.98 per gram of gold and the density 19.3 grams per cubic centimeter, find the cost of:
1. 1 cubic centimeter of gold
2. 1 cubic inch of gold

Solution:

1. $\dfrac{\$5.98}{1 \text{ g}} \times \dfrac{19.3 \text{ g}}{1 \text{ cm}^3} = \dfrac{\$115.414}{1 \text{ cm}^3} = \$115.41/\text{cm}^3$

2. $\dfrac{\$5.98}{1 \text{ g}} \times \dfrac{19.3 \text{ g}}{\text{cm}^3} \times \dfrac{16.39 \text{ cm}^3}{1 \text{ in}^3} = \dfrac{\$1{,}891.6354}{1 \text{ in}^3} = \$1{,}891.64/\text{in}^3$

Complete the problems below. Completed problems appear on page 423.

Sample Set

(a) Convert 1 centimeter per second to inches per minute (nearest 10th).

$$\frac{1 \text{ cm}}{1 \text{ sec}} \times \frac{\boxed{}}{1 \text{ min}} \times \frac{1 \text{ in}}{\boxed{}} = \frac{\boxed{} \text{ in}}{\boxed{} \text{ min}} = \boxed{} \text{ in/min}$$

(b) If the speed of a plane is 880 feet per second, how many kilometers will the plane travel in 1 hour? (Use 1 mile = 1.61 kilometers.)

$$1 \text{ hr} \times \frac{\boxed{}}{1 \text{ hr}} \times \frac{\boxed{}}{1 \text{ min}} \times \frac{880 \text{ ft}}{1 \text{ sec}} \times \frac{1 \text{ mi}}{\boxed{} \text{ ft}} \times \frac{\boxed{}}{1 \text{ mi}}$$

$$= \boxed{} \text{ km}$$

(c) The density of iron is 450 pounds per cubic foot. Find the density in kilograms per cubic meter. (Use 1 cubic meter = 35.3 cubic feet.)

$$\frac{450 \text{ lb}}{1 \text{ ft}^3} \times \frac{\boxed{}}{1 \text{ m}^3} \times \frac{1 \text{ kg}}{\boxed{}} = \frac{\boxed{} \text{ kg}}{\boxed{} \text{ m}^3}$$

$$= \boxed{} \text{ kg/m}^3 \quad \text{nearest whole unit}$$

(d) The density of iron is 450 pounds per cubic foot. Find the weight of 1 cubic inch of iron in ounces.

$$1 \text{ in}^3 \times \frac{1 \text{ ft}^3}{\boxed{}} \times \frac{\boxed{}}{1 \text{ ft}^3} \times \frac{\boxed{}}{1 \text{ lb}} = \boxed{} \text{ oz}$$

(e) If the concentration of morphine sulfate in a stock solution is 2 milligrams per cubic centimeter, how many cubic centimeters of this solution would be needed to have $\frac{1}{15}$ grain. (Use 1 grain = 60 mg.)

$$\frac{1}{15} \text{ grain} \times \frac{\boxed{}}{\boxed{}} \times \frac{\boxed{}}{2 \text{ mg}} = \boxed{} \text{ cm}^3$$

(f) If a grower receives \$110 per ton of peaches, he is receiving how many cents per pound?

$$\frac{\$110}{1 \text{ ton}} \times \frac{1 \text{ ton}}{\boxed{} \text{ lb}} \times \frac{\boxed{} \cent}{\$ \boxed{}} = \frac{\boxed{} \cent}{\boxed{} \text{ lb}} = \boxed{} \cent/\text{lb}$$

(g) The instructions on a box of grass seed state that 1 pound of seed will cover 1,000 square feet of ground. If the seed costs 63 cents per pound, what would be the cost in dollars to cover 500 square meters of ground? (Use 1 square meter = 10.8 square feet.)

$$500 \text{ m}^2 \times \frac{\boxed{}}{1 \text{ m}^2} \times \frac{1 \text{ lb}}{\boxed{}} \times \frac{\boxed{}}{1 \text{ lb}} \times \frac{\boxed{}}{\boxed{}} = \$ \boxed{}$$

Completed Problems

(a) Convert 1 centimeter per second to inches per minute (nearest 10th).

$$\frac{1 \cancel{cm}}{1 \cancel{sec}} \times \frac{\boxed{60 \cancel{sec.}}}{1 \text{ min}} \times \frac{1 \text{ in}}{\boxed{2.54 \cancel{cm}}} = \frac{\boxed{60} \text{ in}}{\boxed{2.54} \text{ min}} = \boxed{23.6} \text{ in/min}$$

(b) If the speed of a plane is 880 feet per second, how many kilometers will the plane travel in 1 hour? (Use 1 mile = 1.61 kilometers.)

$$1 \cancel{hr} \times \frac{\boxed{60 \cancel{min.}}}{1 \cancel{hr}} \times \frac{\overset{10}{\boxed{60 \cancel{sec.}}}}{1 \cancel{min}} \times \frac{\overset{1}{880 \cancel{ft}}}{1 \cancel{sec}} \times \frac{1 \cancel{mi}}{\underset{\underset{1}{6}}{\boxed{5280}} \cancel{ft}} \times \frac{\boxed{1.61 \text{ Km}}}{1 \cancel{mi}}$$

$$= \boxed{966} \text{ km}$$

(c) The density of iron is 450 pounds per cubic foot. Find the density in kilograms per cubic meter. (Use 1 cubic meter = 35.3 cubic feet.)

$$\frac{450 \cancel{lb}}{1 \cancel{ft^3}} \times \frac{\boxed{35.3 \cancel{ft^3}}}{1 \text{ m}^3} \times \frac{1 \text{ kg}}{\boxed{2.2 \cancel{lb}}} = \frac{\boxed{15,885}}{\boxed{2.2}} \frac{\text{kg}}{\text{m}^3}$$

$$= \boxed{7,220} \text{ kg/m}^3 \quad \text{nearest whole unit}$$

(d) The density of iron is 450 pounds per cubic foot. Find the weight of 1 cubic inch of iron in ounces.

$$1 \cancel{in^3} \times \frac{1 \cancel{ft^3}}{\underset{108}{\boxed{1728 \cancel{in^3}}}} \times \frac{\overset{1}{450 \cancel{lb}}}{1 \cancel{ft^3}} \times \frac{\boxed{16 oz.}}{1 \cancel{lb}} = \boxed{4.2} \text{ oz}$$

(e) If the concentration of morphine sulfate in a stock solution is 2 milligrams per cubic centimeter, how many cubic centimeters of this solution would be needed to have $\frac{1}{15}$ grain. (Use 1 grain = 60 mg.)

$$\frac{1}{\underset{1}{\cancel{15}}} \cancel{grain} \times \frac{\overset{\overset{2}{\cancel{4}}}{\boxed{60 \cancel{mg}}}}{1 \cancel{grain}} \times \frac{\boxed{1 cm^3}}{\underset{1}{2 \cancel{mg}}} = \boxed{2} \text{ cm}^3$$

(f) If a grower receives $110 per ton of peaches, he is receiving how many cents per pound?

$$\frac{\$110}{1 \text{ ton}} \times \frac{1 \text{ ton}}{\underset{20}{\cancel{2{,}000}} \text{ lb}} \times \frac{\overset{1}{\cancel{100}} \text{ ¢}}{\cancel{\$} 1} = \frac{110}{20} \text{ ¢/lb} = \boxed{5.5} \text{ ¢/lb}$$

(g) The instructions on a box of grass seed state that 1 pound of seed will cover 1,000 square feet of ground. If the seed costs 63 cents per pound, what would be the cost in dollars to cover 500 square meters of ground? (Use 1 square meter = 10.8 square feet.)

$$\underset{1}{\cancel{500}} \text{ m}^2 \times \frac{\overset{5.4}{\cancel{10.8}} \text{ ft}^2}{1 \text{ m}^2} \times \frac{1 \text{ lb}}{\underset{2}{\underset{1}{\cancel{1000 \text{ ft}^2}}}} \times \frac{63 \text{¢}}{1 \text{ lb}} \times \frac{\$1}{100 \text{¢}} = \$\boxed{3.40}$$

Name: _____

Class: _____

Exercises for Sec. 11-5 Speed

1. Convert 1 centimeter per second to meters per minute.

2. Convert 1 foot per second to yards per minute.

3. Convert 60 miles per hour to yards per second.

4. Convert 80 kilometers per hour to meters per second.

5. Convert 60 miles per hour to kilometers per hour. (Use 1 mile = 1.61 kilometers.)

6. Convert 80 kilometers per hour to miles per hour. (Use 1 kilometer = .62 mile.)

7. Convert 100 feet per minute to meters per second. (Round to the nearest 10th of a meter.)

8. Convert 186,000 miles per second to centimeters per second. (Use 1 mile = 1.61 kilometers; round to the nearest 100,000,000 centimeters.)

9. A plane traveling 400 miles per hour will travel how many kilometers in 2 hours? (Use 1 mile = 1.61 kilometers.)

10. A car traveling 80 kilometers per hour will travel how many miles in 3 hours? Round your answer to the nearest whole mile. (Use 1 kilometer = .62 mile.)

11. A runner averaging 7 meters per second will travel how many miles in 20 minutes? Round your answer to the nearest 100th. (Use 1 kilometer = .62 mile.)

12. A person walking at an average speed of 6 feet per second will travel how many meters in 1 minute. (Use 1 foot = .30 meter.)

13. How many minutes will it take light traveling at a speed of 186,000 miles per second to reach the earth from the sun? The distance from the sun to the earth is 93,000,000 miles. Round your answer to the nearest 10th of a minute.

14. If a particle could travel around the equator of the earth (25,000 miles) at a speed of 186,000 miles per second, how many times would it circle the earth in 1 second? Round your answer to the nearest 10th.
    (*Hint:* Multiply 1 second by conversion factors that cancel second and introduce times. Use the rate 1 time per 25,000 miles.)

Density

15. Convert 450 pounds per cubic foot to pounds per cubic yard.

16. Convert 1 kilogram per cubic meter to grams per cubic centimeter.

17. Convert 1 gram per cubic centimeter to kilograms per cubic meter.

18. Convert 1 ounce per cubic inch to pounds per cubic foot.

19. Convert 100 kilograms per cubic meter to pounds per cubic yard. (Use 1 cubic yard = .76 cubic meters; round to the nearest whole pound.)

20. Convert 500 pounds per cubic foot to kilograms per cubic meter. (Use 1 cubic meter = 35.3 cubic feet; round to the nearest whole kilogram.)

21. Convert 1 ounce per cubic inch to grams per cubic centimeter. (Round to the nearest 10th of a gram.)

22. Convert .01 kilogram per cubic centimeter to pounds per cubic inch. (Round to the nearest 100th of a pound.)

23. The density of copper is 540 pounds per cubic foot. How many pounds of copper are in 1 cubic yard?

24. The density of tin is 7,380 kilograms per cubic meter. How many grams will be in 1 cubic centimeter?

25. The density of silver is 10.5 grams per cubic centimeter. How many pounds are in 1 cubic foot? Round your answer to the nearest pound. (Use 1 cubic foot = .028 cubic meter.)

26. The density of lead is 712 pounds per cubic foot. How many grams are in 1 cubic centimeter? Round your answer to the nearest 10th of a gram. (Use 1 cubic meter = 35.3 cubic feet.)

27. Ebony wood weighs 76 pounds per cubic foot. What is the volume in cubic inches of a carving that weighs 19 pounds?

28. The density of common brick is 125 pounds per cubic foot. What would be the weight in tons of 1,000 bricks if each brick has length = 6 inches, width = 3 inches, and thickness = 2 inches? Round to the nearest 10th of a ton.

Concentration: The equivalent measures that will be used in Exercises 29 through 36 are listed below.

Weight: 60 milligrams = 1 grain

Volume: 1 cubic centimeter = 15 minims or 16 minims (whichever makes cancellation easier)

29. Convert $\frac{1}{4}$ grain per cubic centimeter to milligrams per cubic centimeter.

30. Convert 2 milligrams per cubic centimeter to grains per cubic centimeter.

31. Convert 3 milligrams per cubic centimeter to grains per minim (m).

32. Convert $\frac{1}{12}$ grain per minim (m̥) to milligrams per cubic centimeter.

33. If the concentration of morphine sulfate in a stock solution is 2 milligrams per cubic centimeter, how many cubic centimeters of this solution are needed to have .5 milligram of morphine sulfate?

34. If the concentration of demerol in a stock solution is 40 milligrams per cubic centimeter, how many minims of this solution would be needed to have 30 milligrams of demerol?

35. If the concentration of atropine in a stock solution is 2 milligrams per cubic centimeter, how many cubic centimeters of this solution would be needed to have $\frac{1}{120}$ grain of atropine? Give the answer as a decimal.

36. If the concentration of sodium kiminal in a stock solution is 45 milligrams per cubic centimeter, how many minims of this solution would be needed to have $\frac{1}{2}$ grain of sodium kiminal.
    (*Hint:* Multiply $\frac{1}{2}$ grain by conversion factors that cancel grain and introduce minim.)

Cost per unit

37. Convert 3 cents per inch to dollars per yard.

38. Convert 2 cents per centimeter to dollars per meter.

39. Convert $10,000 per meter to dollars per mile. (Use 1 mile = 1.61 kilometers.)

40. Convert $10 per square yard to dollars per square meter. (Use 1 square meter = 1.2 square yards.)

41. Convert $2 per square meter to dollars per square foot. (Use 1 square foot = .09 square meter.)

42. Convert 50 cents per quart to dollars per cubic foot. (Use 1 cubic foot = 30 quart.)

43. Convert 50 cents per liter to dollars per cubic meter.

44. Convert 60 cents per quart to cents per liter. (Use 1.06 quarts = 1 liter.)

45. Convert 5 cents per ounce to dollars per pound.

46. Convert 10 cents per gram to dollars per kilogram.

47. Convert $10 per pound to dollars per kilogram.

48. Convert $220 per kilogram to dollars per pound.

49. How much would 40 inches of material cost if the material costs $5 per yard?

50. How much would 1 kilogram of gold cost if the price of gold is $4.50 per gram?

51. What would be the cost of the soybeans needed to fill a storage bin with a volume of 10,000 cubic feet if the price of soybeans is $7.60 per bushel. (Use 1 bushel = 1.24 cubic feet. Round to the nearest dollar.)

52. If gold is selling for $180 per ounce, how many pounds of gold could you buy with $4,500?

53. If platinum is selling for $200 per ounce, what would 10 cubic inches of platinum cost? (The density of platinum is 12.4 ounces per cubic inch.)

54. How much would it cost to paint 10 square yards of wall if paint costs $6 per gallon and 1 gallon will cover 600 square feet?

55. A rectangle surface 10 inches by 20 inches is to be plated .05 inches thick on one side with silver. If silver costs $5.80 per ounce, how much will it cost to plate the surface? (The density of silver is 6.05 ounces per cubic inch.)

Name: _____

Class: _____

Unit 11 Test
(Sections 1 through 4)

1. Convert 7,500,000 milligrams to kilograms.

2. Convert .375 kilometer to centimeters.

3. Convert 1,000,000 cubic millimeters to cubic meters.

4. Convert 3,572 cubic centimeters to liters.

5. A prism has length = 500 centimeters, width = 200 centimeters, and height = 300 centimeters. Find the volume in cubic meters.

6. Convert $\frac{3}{4}$ pounds to ounces.

7. Convert 220 yards to miles.

8. Convert 1 cubic foot to gallons (nearest 10th of a gallon).

9. A rectangle has length = 81 inches and width = 16 inches. Find the area in square feet.

10. Convert 100 yards to meters (nearest 10th of a meter).

11. Convert 250 kilograms to pounds (nearest whole pound).

12. Convert 1 gallon to liters (nearest 10th of a liter).

13. Convert 22 ounces to grams.

14.   5 yd   2 ft   10 in   15.   8 kg   375 g   460 mg
       2 yd   1 ft   11 in   7 kg   729 g   782 mg
     + 3 yd   1 ft   9 in   + 6 kg   875 g   231 mg

16. 5 m   30 cm   8 mm
    −2 m   69 cm   9 mm

17. 3 liters   250 cm³
    −1 liter   500 cm³

18. If 9 ounces of beans cost 12 cents, how much will 15 ounces cost?

19. How many pounds of ore would be needed to produce 200 pounds of pure copper if there are 160 pounds of pure copper in 2,000 pounds of ore?

20. What would it cost to drive a car 300 miles if gasoline costs 60 cents per gallon and the car can be driven 18 miles on 1 gallon of gasoline?

Name: _____

Class: _____

Unit 11 Test
(Section 5)

1. Convert 150 miles per hour to yards per minute.

2. Convert 600 kilometers per hour to centimeters per second.

3. Convert 10 feet per minute to centimeters per second.

4. A car traveling 66 feet per second will travel how many miles in 3 hours?

5. A car travels 25 meters per second. How many hours will it take the car to travel 900 kilometers?

6. Convert 4 ounces per cubic inch to pounds per cubic foot.

7. Convert 1,000 kilograms per cubic meter to pounds per cubic inch. Round to the nearest 100th of a pound.

8. Convert .01 kilogram per cubic centimeter to pounds per cubic inch. Round to the nearest 100th of a pound.

9. The density of platinum is 21.5 grams per cubic centimeter. How much do 5 cubic centimeters of platinum weigh?

10. The density of platinum is 12.4 ounces per cubic inch. What would be the volume in cubic inches of 62 ounces of platinum?

11. Convert 5 grams per liter to milligrams per cubic centimeter.

12. Convert 5 milligrams per cubic centimeter to grains per cubic centimeter. (1 grain = 60 milligrams.)

13. Convert $\frac{1}{12}$ grain per cubic centimeter to milligrams per minim. (1 grain = 60 milligrams; 1 cubic centimeter = 15 m︎.)

14. The concentration of a solution of morphine sulfate is 6 milligrams per cubic centimeter. How many cubic centimeters of this solution would be needed to have 3 milligrams of morphine sulfate?

15. The concentration of a solution of atropine is 3 milligrams per cubic centimeter. How many minims of this solution would be needed to have 2 milligrams of atropine?

16. Convert $5 per pound to cents per ounce.

17. Convert $2,500 per cubic meter to cents per cubic centimeter.

18. Convert 5 cents per inch to cents per centimeter. Round to the nearest cent.

19. If platinum is selling for $200 per ounce, how much would 2 pounds of platinum cost?

20. If platinum is selling for $200 per ounce, how many cubic inches could you buy with $24,800? The density of platinum is 12.4 ounces per cubic inch. Round to the nearest 10th.

Section 12-1 Positive Exponents

A product in which the same number is repeated as a factor may be written in **exponential form**.

Exponential Form

$$(3)(3) = 3^2 \quad \textbf{3 squared}$$
$$(3)(3)(3) = 3^3 \quad \textbf{3 cubed}$$
$$(10)(10)(10)(10) = 10^4 \quad \textbf{10 to the fourth power}$$
$$(5)(5)(5)(5)(5)(5) = 5^6 \quad \textbf{5 to the sixth power}$$

The exponential form 5^6 is called a **power** of 5. It is read "5 to the sixth power" or "the sixth power of 5." The number 5 is called the **base** of the power, and the number 6 is called the **exponent**. Note that the exponent counts the number of times the base is used as a factor.

A single number is understood to have an exponent of 1. For example, $3 = 3^1$ and $10 = 10^1$.

Multiplication of Powers

When powers with the same base are multiplied, the exponents may be added. For example,

$$(3^3)(3^5) = 3^{3+5} = 3^8$$

The justification for doing this is demonstrated below.

$$(3^3)(3^5) = [\underbrace{(3)(3)(3)}_{\text{3 threes grouped}}][\underbrace{(3)(3)(3)(3)(3)}_{\text{5 threes grouped}}]$$
$$= \underbrace{(3)(3)(3)(3)(3)(3)(3)(3)}_{\text{8 threes (any order)}}$$
$$= 3^8$$

Note that the justification depends on the associative law of multiplication.

The rule for multiplying like bases is given in general form below.

$$\boxed{a^m a^n = a^{m+n}}$$

Examples

Write each product as a single power of the base.

(a) $(5^3)(5^2) = 5^5$
(b) $(10^{32})(10^{53}) = 10^{85}$

Write each product as a product in which each base appears once.

(c) $(2^3)(3^5)(2^8)(3^2) = [(2^3)(2^8)][(3^5)(3^2)] = (2^{11})(3^7)$

(d) $(7^2)(5^3)(5^2)(3^3)(7^4)(3) = (3^4)(5^5)(7^6)$
(*Note:* $3 = 3^1$)

In general, you may not multiply bases.

$$(3^2)(3^5) \neq 9^7$$
$$(3^2)(5^4) \neq 15^6$$

The **only time** bases may be multiplied is when the exponents in each factor are the same. In this case, the exponent in the product will be the exponent of the factors.

$$a^n b^n = (ab)^n$$

Example

same exponent
$(3^2)(5^2) = 15^2$

Reasoning

$(3^2)(5^2) = [(3)(3)][(5)(5)]$
$= [(3)(5)][(3)(5)]$
$= [15][15]$
$= 15^2$

Division of Powers When powers with like bases are divided, the exponents may be subtracted.

Examples

$$\frac{5^7}{5^3} = \frac{5^{7-3}}{1} = \frac{5^4}{1} = 5^4$$

$$\frac{4^2}{4^6} = \frac{1}{4^{6-2}} = \frac{1}{4^4}$$

Reasoning

$$\frac{5^7}{5^3} = \frac{\cancel{(5)}\cancel{(5)}\cancel{(5)}(5)(5)(5)(5)}{\cancel{(5)}\cancel{(5)}\cancel{(5)}} = \frac{5^4}{1}$$

$$\frac{4^2}{4^6} = \frac{\cancel{(4)}\cancel{(4)}}{\cancel{(4)}\cancel{(4)}(4)(4)(4)(4)} = \frac{1}{4^4}$$

In each case, the smaller exponent was subtracted from the larger. It is possible to subtract a larger exponent from a smaller exponent. The result is a negative exponent. Negative exponents will be covered in the next section.

The general rule for dividing powers with like bases is stated below.

$$\frac{a^m}{a^n} = a^{m-n} \qquad a \neq 0$$

$$\frac{a^m}{a^n} = \frac{1}{a^{n-m}} \qquad a \neq 0$$

Examples

Find each quotient.

(a) $\dfrac{7^5}{7^3} = 7^2 = 49$ $7^2 = (7)(7) = 49$

(b) $\dfrac{10^{85}}{10^{88}} = \dfrac{1}{10^3} = \dfrac{1}{1{,}000}$ $10^3 = (10)(10)(10) = 1{,}000$

(c) $\dfrac{(5^3)(3^6)}{(5^5)(3^2)} = \left(\dfrac{5^3}{5^5}\right)\left(\dfrac{3^6}{3^2}\right) = \left(\dfrac{1}{5^2}\right)\left(\dfrac{3^4}{1}\right)$

$\qquad = \dfrac{3^4}{5^2} = \dfrac{81}{25}$ $3^4 = (3)(3)(3)(3) = 81$
$5^2 = (5)(5) = 25$

(d) $\dfrac{(8^6)(5^4)}{(5^3)(8^8)} = \dfrac{5^1}{8^2} = \dfrac{5}{64}$ $8^2 = (8)(8) = 64$

Complete each problem below. Completed problems appear on page 447.

Sample Set

(a) $(9^{12})(9^5) = 9^{\boxed{}}$ (b) $(3^4)(5^7)(3^5)(5^3) = (3^{\boxed{}})(5^{\boxed{}})$

(c) $\dfrac{9^{14}}{9^{12}} = 9^{\boxed{}} = \boxed{}$ (d) $\dfrac{3^5}{3^8} = \dfrac{1}{3^{\boxed{}}} = \dfrac{1}{\boxed{}}$

(e) $\dfrac{(4^5)(2^7)}{(4^4)(2^5)} = (4^{\boxed{}})(2^{\boxed{}}) = (\ \ )(\ \ ) = \boxed{}$

(f) $\dfrac{(2^3)(3^5)}{(2^6)(3^7)} = \dfrac{1}{(2^{\boxed{}})(3^{\boxed{}})} = \dfrac{1}{(\ \ )(\ \ )} = \dfrac{1}{\boxed{}}$

(g) $\dfrac{(3^5)(2^8)}{(3^7)(2^5)} = \dfrac{2^{\boxed{}}}{3^{\boxed{}}} = \dfrac{\boxed{}}{\boxed{}}$

The Tower of Hanoi

The Tower of Hanoi game consists of three pegs, with disks of different sizes set on one of the pegs. The largest disk is on the bottom, the next largest disk is on that one, and so on up to the smallest disk, which is on the top. The object is to transfer all the disks to one of the other pegs without placing a larger disk on top of a smaller one.

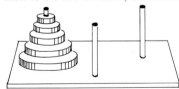

The number of transfers required to move one disk is clearly 1. The number of transfers required to move two disks is $3 = 2^2 - 1$. The number for three disks is $7 = 2^3 - 1$. In general, the number of transfers required to move n disks is $2^n - 1$. You can check this for yourself by placing four or five books of different sizes in a pile and then transferring the books from pile to pile.

There is a myth that, at the time of creation, priests in the temple at Benares started to transfer 64 disks in the manner described. When they finished, the world was to come to an end. If they transferred one disk per second, never making a mistake, it would then take them $2^{64} - 1$ seconds.

If you were to compute $2^{64} - 1$ correctly (try it), you would find that

$$2^{64} - 1 = 18{,}446{,}744{,}073{,}709{,}551{,}615$$

Converting this many seconds into years, one finds that the time required for the priests to transfer all 64 disks would be slightly more than

$$5{,}845{,}420{,}460{,}906 \text{ years}$$

This number of years is approximately 1,000 times the estimated age of the earth. You can see that there would be no need to worry.

A similar example is provided by another myth. A story is told that a king had a very grave problem to solve. The court jester found the solution, and the king was so pleased that he promised the jester anything he wanted, even half his kingdom. The jester, being a practical joker, asked the king to give him one grain of wheat on a square of a checker board, two grains on a second square, four grains on a third square, eight on a fourth, and so on, doubling the number of grains until all 64 squares were covered. The king was delighted at such a modest request—that is, until he tried to fill the board. How much wheat would be required? The table below shows the number of grains.

Number of Squares Filled	Total Number of Grains
1	1
2	$1 + 2 = 3 = 2^2 - 1$
3	$1 + 2 + 4 = 7 = 2^3 - 1$
4	$1 + 2 + 4 + 8 = 15 = 2^4 - 1$
...	...
64	$2^{64} - 1$

Someone has estimated that $2^{64} - 1$ grains of wheat would cover the entire earth to a depth of 6 inches.

These examples demonstrate how quickly powers of 2 increase.

Completed Problems

(a) $(9^{12})(9^5) = 9^{\boxed{17}}$

(b) $(3^4)(5^7)(3^5)(5^3) = (3^{\boxed{9}})(5^{\boxed{10}})$

(c) $\dfrac{9^{14}}{9^{12}} = 9^{\boxed{2}} = \boxed{81}$

(d) $\dfrac{3^5}{3^8} = \dfrac{1}{3^{\boxed{3}}} = \dfrac{1}{\boxed{27}}$

(e) $\dfrac{(4^5)(2^7)}{(4^4)(2^5)} = (4^{\boxed{1}})(2^{\boxed{2}}) = (4)(4) = \boxed{16}$

(f) $\dfrac{(2^3)(3^5)}{(2^6)(3^7)} = \dfrac{1}{(2^{\boxed{3}})(3^{\boxed{2}})} = \dfrac{1}{(\boxed{8})(\boxed{9})} = \dfrac{1}{\boxed{72}}$

(g) $\dfrac{(3^5)(2^8)}{(3^7)(2^5)} = \dfrac{2^{\boxed{3}}}{3^{\boxed{2}}} = \dfrac{\boxed{8}}{\boxed{9}}$

Name: _____

Class: _____

Exercises for Sec. 12-1 Write each product as a single power of the base.

1. $(5^7)(5^3) =$

2. $(3^7)(3^9) =$

3. $(8^{35})(8^{21}) =$

4. $(6^2)(6^3)(6^7) =$

5. $(4^3)(4^5)(4^2) =$

6. $(7^{13})(7^{23})(7^{15}) =$

Write each product as a product in which each base appears once.

7. $(2^5)(3^2)(3^7)(2^3) =$

8. $(5^2)(4^3)(5^7)(4^5) =$

9. $(2^9)(7^5)(2^{10})(7^3) =$

10. $(7^3)(8^2)(8^{12})(7^4) =$

11. $(3^5)(2^7)(5^3)(2^6)(3^3)(2^8) =$

12. $(6^5)(8^3)(6^2)(5^9)(8^4)(5^4) =$

Compute each quotient.

13. $\dfrac{5^5}{5^3} =$

14. $\dfrac{4^7}{4^4} =$

15. $\dfrac{6^5}{6^8} =$

16. $\dfrac{8^{10}}{8^{12}} =$

17. $\dfrac{3^6}{3^2} =$

18. $\dfrac{2^3}{2^8} =$

19. $\dfrac{(3^5)(2^7)}{(3^4)(2^5)} =$

20. $\dfrac{(5^3)(4^5)}{(5^1)(4^4)} =$

21. $\dfrac{(7^2)(3^2)}{(7^3)(3^4)} =$

22. $\dfrac{(3^3)(7^5)}{(3^5)(7^6)} =$

23. $\dfrac{(5^4)(2^5)}{(5^2)(2^8)} =$

24. $\dfrac{(3^2)(8^8)}{(3^5)(8^6)} =$

25. $\dfrac{(4^2)(2^7)}{(2^3)(4^5)} =$

26. $\dfrac{(6^3)(4^5)}{(4^2)(6^5)} =$

27. $\dfrac{(6^2)(9^7)}{(9^5)(6^5)} =$

28. $\dfrac{(2^5)(9^9)}{(9^7)(2^{10})} =$

29. $\dfrac{(2^5)(3^3)(5^4)}{(5^6)(2^3)(3^2)} =$

30. $\dfrac{(3^6)(4^7)(2^7)}{(4^5)(2^{11})(3^5)} =$

Section 12-2 Negative Exponents

It is desirable to have the rules for multiplying and dividing powers apply to all choices of exponents. For example, the rule

$$\frac{a^m}{a^n} = a^{m-n} \qquad a \neq 0$$

should apply in the case when $m = 2$ and $n = 5$.

$$\frac{a^2}{a^5} = a^{2-5} = a^{-3}$$

In the previous section, you learned that

$$\frac{a^2}{a^5} = \frac{1}{a^3}$$

The power a^{-3} is, therefore, defined to be $1/a^3$. In general,

$$a^{-n} = \frac{1}{a^n}$$

$$a^n = \frac{1}{a^{-n}}$$

What should the quotient be when $n = m$?

$$\frac{a^m}{a^m} = a^{m-m} = a^0$$

Since the numerator and denominator are equal, the fraction must be 1.

$$\frac{a^m}{a^m} = 1$$

The power a^0 is, therefore, defined to be 1.

$$a^0 = 1$$

The definitions are consistent with the multiplication rule

$$a^m a^n = a^{m+n}$$

For example,

$$(3^5)(3^{-2}) = 3^{5+(-2)} = 3^{5-2} = 3^3$$

agrees with

$$(3^5)(3^{-2}) = (3^5)\left(\frac{1}{3^2}\right) = \frac{3^5}{3^2} = 3^3$$

452 Exponents and Roots

As a second example,
$$(5^{-3})(5^3) = 5^{-3+3} = 5^0 = 1$$

agrees with
$$(5^{-3})(5^3) = \left(\frac{1}{5^3}\right)(5^3) = \frac{5^3}{5^3} = 1$$

Important Point The zero exponent does not count the number of times the base number is being "multiplied by itself." a^0 **is defined to be 1** so that the multiplication and division rules will be consistent for all exponents.

Warning: A Common Error Students often assume that a negative exponent leads to a negative answer. Note that

$$5^{-4} = \frac{1}{5^4} = \frac{1}{625} \quad \text{is positive}$$

The negative exponent simply indicates the power 5^{-4} is to be inverted and the exponent changed in sign.

Examples

Simplify each product so that the base appears once with a positive exponent.

(a) $(3^5)(3^{-7}) = 3^{5-7} = 3^{-2} = \dfrac{1}{3^2}$ answer

(b) $(5^{-2})(5^{-4}) = 5^{-2-4} = 5^{-6} = \dfrac{1}{5^6}$

(c) $(4^{-3})(4^7) = 4^{-3+7} = 4^4$

(d) $(2^{-5})(2^8)(2^{-6}) = 2^{-5+8-6} = 2^{-3} = \dfrac{1}{2^3}$

Sample Problem 12-1 Write $5^{-2}/5^3$ as a fraction in which only a positive exponent appears.

Solution: Method 1. Using $a^m/a^n = a^{m-n}$,

$$\frac{5^{-2}}{5^3} = 5^{-2-3} = 5^{-5} = \frac{1}{5^5} \quad \text{answer}$$

Method 2. Using $a^m/a^n = 1/a^{n-m}$,

$$\frac{5^{-2}}{5^3} = \frac{1}{5^{3-(-2)}} = \frac{1}{5^{3+2}} = \frac{1}{5^5}$$

Method 3. Using $a^{-n} = 1/a^n$,

$$\frac{5^{-2}}{5^3} = (5^{-2})\left(\frac{1}{5^3}\right) = \left(\frac{1}{5^2}\right)\left(\frac{1}{5^3}\right) = \frac{1}{5^5}$$

Method 3 may actually be the easiest. See the explanation that follows.

Sec. 12-2 Negative Exponents

The equalities

$$\frac{a^{-n}}{1} = \frac{1}{a^n}$$

and

$$\frac{a^n}{1} = \frac{1}{a^{-n}}$$

tell us that powers in a term may be shifted from numerator to denominator or denominator to numerator. All one has to do is change the sign of the exponent.

Examples Simplify each fraction so that the base appears once with a positive exponent.

(a) $\dfrac{5^{-2}}{5^3} = \dfrac{1}{(5^2)(5^3)} = \dfrac{1}{5^5}$

(b) $\dfrac{3^4}{3^{-6}} = (3^4)(3^6) = 3^{10}$

(c) $\dfrac{4^{-2}}{4^{-5}} = (4^{-2})(4^5) = 4^3$

(d) $\dfrac{2^{-7}}{2^{-5}} = \dfrac{1}{(2^7)(2^{-5})} = \dfrac{1}{2^2}$

Please see the material on page 454 before completing the sample set below.

Complete the problems below. Completed problems appear on page 455.

Sample Set

(a) $(5^{15})(5^{-8}) = 5^{\Box}$

(b) $(8^{-7})(8^3) = 8^{\Box} = \dfrac{1}{8^{\Box}}$

(c) $(7^4)(7^{-5})(7^{-3}) = 7^{\Box} = \dfrac{1}{7^{\Box}}$

(d) $\dfrac{6^{-4}}{6^3} = \dfrac{1}{(6^{\Box})(6^{\Box})} = \dfrac{1}{6^{\Box}}$

(e) $\dfrac{9^{-5}}{9^{-8}} = (9^{-5})(9^{\Box}) = 9^{\Box}$

(f) $\dfrac{(2^{-3})(5^2)}{(2^{-7})(5^4)} = \dfrac{(2^{\Box})(2^{-3})}{(5^4)(5^{\Box})} = \dfrac{2^{\Box}}{5^{\Box}} = \dfrac{\Box}{\Box}$

(g) $\dfrac{(3^{-2})(2^{-2})}{(3^2)(2^{-5})} = \dfrac{(2^{-2})(2^{\Box})}{(3^{\Box})(3^2)} = \dfrac{2^{\Box}}{3^{\Box}} = \dfrac{\Box}{\Box}$

(h) $\dfrac{(2^{-3})(3^2)(2^{-2})}{(3^{-3})(2^{-7})(3^6)} = \dfrac{(2^{\Box})(2^{-3})(2^{-2})}{(3^{\Box})(3^{-3})(3^6)} = \dfrac{2^{\Box}}{3^{\Box}} = \dfrac{\Box}{\Box} = \Box$

How do you know in advance whether to "shift" a power up or down? It really makes no difference which way you do it. For example,

$$\frac{2^{-7}}{2^{-5}} = (2^{-7})(2^5) = 2^{-2} = \frac{1}{2^2}$$

yields the same result as

$$\frac{2^{-7}}{2^{-5}} = \frac{1}{(2^7)(2^{-5})} = \frac{1}{2^2}$$

Examples Compute each quotient.

(a) $\quad \dfrac{(5^{-2})(3^3)}{(5^2)(3^{-2})} = \dfrac{(3^3)(3^2)}{(5^2)(5^2)} = \dfrac{3^5}{5^4} = \dfrac{243}{625}$

or $\quad \dfrac{(5^{-2})(3^4)}{(5^2)(3^{-2})} = \dfrac{(5^{-2})(5^{-2})}{(3^{-4})(3^{-2})} = \dfrac{5^{-4}}{3^{-6}} = \dfrac{3^6}{5^4} = \dfrac{243}{625}$

(*Note:* $3^5 = (3)(3)(3)(3)(3) = 243$; $5^4 = (5)(5)(5)(5) = 625$)

(b) $\quad \dfrac{(10^{-5})(8^{-3})}{(8^{-4})(10^{-3})} = \dfrac{(8^4)(8^{-3})}{(10^5)(10^{-3})} = \dfrac{8^1}{10^2} = \dfrac{8}{100}$

or $\quad \dfrac{(10^{-5})(8^{-3})}{(8^{-4})(10^{-3})} = \dfrac{(10^{-5})(10^3)}{(8^{-4})(8^3)} = \dfrac{10^{-2}}{8^{-1}} = \dfrac{8^1}{10^2} = \dfrac{8}{100}$

(c) $\quad \dfrac{(7^{-4})(3^{-5})}{(3^{-5})(7^{-6})} = \dfrac{(7^{-4})(7^6)}{(3^{-5})(3^5)} = \dfrac{7^2}{3^0} = \dfrac{7^2}{1} = 7^2 = 49$

(d) $\quad \dfrac{(2^{-3})(3^4)(2^{-4})}{(2^{-9})(3^{-4})(3^9)}$

Method 1. Multiply powers with like bases first; then shift:

$$\dfrac{(2^{-3})(3^4)(2^{-4})}{(2^{-9})(3^{-4})(3^9)} = \dfrac{(2^{-7})(3^4)}{(2^{-9})(3^5)} = \dfrac{(2^9)(2^{-7})}{(3^{-4})(3^5)} = \dfrac{2^2}{3^1} = \dfrac{4}{3}$$

Method 2. Shift powers; then multiply:

$$9 - 3 - 4 = 2$$
$$\dfrac{(2^{-3})(3^4)(2^{-4})}{(2^{-9})(3^{-4})(3^9)} = \dfrac{(2^9)(2^{-3})(2^{-4})}{(3^{-4})(3^{-4})(3^9)} = \dfrac{2^2}{3^1} = \dfrac{4}{3}$$
$$-4 - 4 + 9 = 1$$

Warning Powers may not be shifted from numerator to denominator, or vice versa, when there are addition or subtraction operations in the fraction. For example,

$$\frac{3^{-2} + 2^3}{2^{-5} - 3^4} \neq \frac{2^5 + 2^3}{3^2 - 3^4}$$

Completed Problems

(a) $(5^{15})(5^{-8}) = 5^{\boxed{7}}$

(b) $(8^{-7})(8^3) = 8^{\boxed{-4}} = \dfrac{1}{8^{\boxed{4}}}$

(c) $(7^4)(7^{-5})(7^{-3}) = 7^{\boxed{-4}} = \dfrac{1}{7^{\boxed{4}}}$

(d) $\dfrac{6^{-4}}{6^3} = \dfrac{1}{(6^{\boxed{4}})(6^{\boxed{3}})} = \dfrac{1}{6^{\boxed{7}}}$

(e) $\dfrac{9^{-5}}{9^{-8}} = (9^{-5})(9^{\boxed{8}}) = 9^{\boxed{3}}$

(f) $\dfrac{(2^{-3})(5^2)}{(2^{-7})(5^4)} = \dfrac{(2^{\boxed{7}})(2^{-3})}{(5^4)(5^{\boxed{-2}})} = \dfrac{2^{\boxed{4}}}{5^{\boxed{2}}} = \boxed{\dfrac{16}{25}}$

(g) $\dfrac{(3^{-2})(2^{-2})}{(3^2)(2^{-5})} = \dfrac{(2^{-2})(2^{\boxed{5}})}{(3^{\boxed{2}})(3^2)} = \dfrac{2^{\boxed{3}}}{3^{\boxed{4}}} = \boxed{\dfrac{8}{81}}$

(h) $\dfrac{(2^{-3})(3^2)(2^{-2})}{(3^{-3})(2^{-7})(3^6)} = \dfrac{(2^{\boxed{7}})(2^{-3})(2^{-2})}{(3^{\boxed{-2}})(3^{-3})(3^6)} = \dfrac{2^{\boxed{2}}}{3^{\boxed{1}}} = \dfrac{\boxed{4}}{3} = \boxed{1\tfrac{1}{3}}$

Name: _____

Class: _____

Exercises for Sec. 12-2 Simplify each product so that the base appears once with a positive exponent.

1. $(3^7)(3^{-15}) =$
2. $(5^8)(5^{-20}) =$
3. $(4^{12})(4^{-8}) =$
4. $(6^{17})(6^{-5}) =$
5. $(5^{-7})(5^{-6}) =$
6. $(7^{-3})(7^{-8}) =$
7. $(8^{-2})(8^7)(8^{-10}) =$
8. $(6^{-8})(6^{-4})(6^{-5}) =$
9. $(2^{-6})(2^{18})(2^{-7})(2^{-6}) =$
10. $(9^{-7})(9^{10})(9^9)(9^{-12}) =$

Simplify each fraction so that the base appears once with a positive exponent.

11. $\dfrac{5^{-2}}{5^3} =$
12. $\dfrac{3^4}{3^{-5}} =$
13. $\dfrac{10^{10}}{10^{-7}} =$
14. $\dfrac{6^{-5}}{6^7} =$
15. $\dfrac{2^{-8}}{2^{-5}} =$
16. $\dfrac{2^{-2}}{2^{-5}} =$
17. $\dfrac{7^{-7}}{7^{-13}} =$
18. $\dfrac{9^{-8}}{9^{-3}} =$

Compute each quotient.

19. $\dfrac{(2^{-2})(3^2)}{(2^2)(3^{-1})} =$

20. $\dfrac{(5^{-2})(3^1)}{(3^{-3})(5^1)} =$

21. $\dfrac{(5^{-2})(7^{-3})}{(5^1)(7^{-5})} =$

22. $\dfrac{(3^{-2})(2^{-2})}{(3^{-3})(2^3)} =$

23. $\dfrac{(2^{-3})(3^{-5})}{(2^{-5})(3^{-2})} =$

24. $\dfrac{(2^{-4})(7^{-7})}{(7^{-5})(2^{-8})} =$

25. $\dfrac{(2^{-2})(4^{-6})(2^5)}{(2^3)(4^{-5})(4^1)} =$

26. $\dfrac{(3^{-5})(7^5)(7^{-8})}{(3^{-4})(7^{-3})(3^{-2})} =$

27. $\dfrac{(5^{-2})(4^{-5})(4^8)}{(4)(5^3)(5^{-7})} =$

28. $\dfrac{(8^{-3})(3^{-2})(8^{-6})}{(3^4)(3^{-3})(8^{-11})} =$

29. $\dfrac{(7^{-5})(3^2)(3^{-5})(5^{-5})}{(3^{-8})(5^{-3})(3^4)(7^{-7})} =$

30. $\dfrac{(2^{-8})(5^3)(4^{-2})(2^2)}{(2^{-3})(5^5)(5^{-1})(4^{-4})} =$

Section 12-3 Scientific Notation

It is not possible to measure the distance to a star, the weight of an atom, or even a person's weight with absolute precision. The precision of a measurement depends on the measuring instrument being used. For example, if you weigh yourself on a bathroom scale, you might be able to find your weight only to the nearest ½ pound. The number one uses to represent a measurement is, therefore, an approximation.

Measurements that occur in nature are often exceedingly large or very small.

Examples of Large Measurements

1. The distance light travels in 1 year is approximately

    5,900,000,000,000 miles = 5 trillion 900 billion miles

2. A galaxy is a large cluster of stars. The number of stars in the Andromeda galaxy is estimated to be

    100,000,000,000 stars = 100 billion stars

3. The Milky Way galaxy, to which our sun belongs, has an estimated diameter of

    590,000,000,000,000,000 miles

4. When a star grows very old, it collapses and becomes small and very dense. When a medium-size star collapses, it becomes a white dwarf. The density of a typical white dwarf might be

    1,000,000 grams per cubic centimeter

Examples of Small Measurements

5. The radius of a hydrogen atom is approximately

    .0000000053 centimeter

6. The mass of a proton, which is a particle in the nucleus of an atom, is approximately

    .00000000000000000000000167 gram
    23 zeros

Some numbers are rounded simply for convenience. For example, the distance light travels in 1 year could have been expressed more accurately. In other cases, a number may express a measurement to the best known precision. The radius of a hydrogen atom is an example.

Significant Digits, Accuracy, and Precision

Recall that every measurement is an approximation. Let us assume that the approximation 6,800,000 pounds is the result of rounding to the nearest 100,000 pounds. A more accurate approximation might have been 6,782,000 pounds. The entire **accuracy** of the measurement is then expressed by the digits 6 and 8, and they are there called **significant digits.**

The zeros to the right of the significant digits will be referred to as **placement zeros.**

6,<u>8</u>00,000   significant digits underlined

As another example, suppose the approximation 50,300 is rounded to the nearest 1,000. The result would be <u>50</u>,000. The accuracy of the measurement is now expressed by the zero in the thousands' place as well as the 5 in the ten-thousands' place. Therefore, both digits are significant digits.

The above examples demonstrate that one cannot tell by looking at a whole-number approximation ending in zeros, which of these zeros are significant. One would have to know the place to which the number was rounded. This is not the case for approximations which are less than 1. For example, when one writes .00<u>4</u> gram, it is generally understood that the approximation is rounded to the last place in the decimal. In this case, the digit 4 is significant. The zeros to the left of 4 are placement zeros. As another example, if the approximation .03964 is rounded to the nearest 1,000th, the result is .0<u>40</u>. Here the zero in the thousandths' place expresses part of the accuracy of the measurement and is a significant digit.

Below are further examples of approximations with the significant digits underlined.

Approximation	Rounded to the
<u>85</u>,000,000	nearest 1,000,000
<u>5,00</u>0,000	nearest 10,000
<u>3,005</u>,000	nearest 1,000
<u>3.05</u>	nearest 100th
<u>2.070</u>	nearest 1,000th
.000<u>8</u>	nearest 10,000th
.000<u>90</u>	nearest 100,000th

> **The accuracy of an approximation is determined by the number of significant digits in the approximation**

The **precision** of a measurement is expressed as the smallest unit in which the measurement is made. For example, if a measurement has been determined to be 3.265 grams, the precision is .001 gram, the unit in the last place.

Which of the measurements below is more precise? Which is more accurate?

.006 gram   or   3.51 grams

Here, .006 gram is more precise, since the measurement is to the nearest

1,000th, a smaller unit than appears in 3.51 grams. However, 3.51 grams has more significant digits and is, therefore, more accurate.

If a better measuring instrument is used and the measurement is determined to be .0062, instead of .006, the measurement is now more precise and more accurate.

Scientific Notation Notice that all the measurements given at the beginning of this section have a large number of placement zeros. It is convenient to represent such numbers using **scientific notation.**

$$\underbrace{530{,}000}_{5 \text{ places}} = 5.3 \times \underbrace{100{,}000}_{5 \text{ zeros}} = \underbrace{5.3 \times 10^5}_{\text{exponent} = 5}$$

$$\underbrace{198{,}000{,}000{,}000}_{11 \text{ places}} = 1.98 \times \underbrace{100{,}000{,}000{,}000}_{11 \text{ zeros}} = \underbrace{1.98 \times 10^{11}}_{\text{exponent} = 11}$$

The exponent in the scientific notation for a large number equals the number of places the decimal point is shifted to the left.

$$\underbrace{.0007}_{4 \text{ places}} = \frac{7}{\underbrace{10{,}000}_{4 \text{ zeros}}} = \frac{7}{10^4} = \underbrace{7 \times 10^{-4}}_{\text{exponent} = -4}$$

$$\underbrace{.0000062}_{6 \text{ places}} = \frac{6.2}{\underbrace{1{,}000{,}000}_{6 \text{ zeros}}} = \frac{6.2}{10^6} = \underbrace{6.2 \times 10^{-6}}_{\text{exponent} = -6}$$

The exponent in the scientific notation for a small number is the negative of the number of places the decimal point is shifted to the right.

How far do you shift the decimal point? It is customary to shift the decimal so that one nonzero digit is to the left of the decimal point; that is, a number N is expressed in scientific notation when $N = K \times 10^p$, where $1 \leq K < 10$ and p is a positive or negative integer. The number K is called the **coefficient** of the scientific notation.

The numbers given at the beginning of this section are written below in scientific notation with the significant digits underlined.

1. $\underline{5.9} \times 10^{12}$ miles
2. $\underline{1} \times 10^{11}$ stars
3. $\underline{5.9} \times 10^{17}$ miles
4. $\underline{1.0} \times 10^6$ grams per cubic centimeter
5. $\underline{5.3} \times 10^{-9}$ centimeter
6. $\underline{1.67} \times 10^{-24}$ gram

Note that the significant digits in each number are contained in the coefficient of the scientific notation. When numbers are multiplied, the product should be rounded to the least number of significant digits appearing in any factor.

Examples

(a) $(2.8 \times 10^{-7})(\underline{3} \times 10^5)$
$= 2.8 \times \underline{3} \times 10^{-7} \times 10^5$
$= 8.4 \times 10^{-2}$
$= \underline{8} \times 10^{-2}$

the number $\underline{3}$ has only one significant digit; the product 8.4 was, therefore, rounded to $\underline{8}$, with one significant digit

(b) $\underline{37}0,000 \times \underline{48},000$
$= \underline{3.7} \times 10^5 \times \underline{4.8} \times 10^4$
$= 17.76 \times 10^9$
$= \underline{18} \times 10^9$

both $\underline{3.7}$ and $\underline{4.8}$ have two significant digits; the product 17.76 was, therefore, rounded to $\underline{18}$ with two significant digits

Why do we use this rounding technique? Recall that each measurement is an approximation. Products of measurements are, therefore, approximations and cannot have more accuracy than the least accurate measurement. This accuracy is expressed by the number of significant digits. **The product is, therefore, rounded to the same number of significant digits as appear in the least accurate measurement.**

The same rounding rule holds when dividing.

Examples

(a) $\dfrac{17.34 \times 10^{-5}}{\underline{6.8} \times 10^{-3}} = \dfrac{17.34}{\underline{6.8}} \times \dfrac{10^{-5}}{10^{-3}}$
$= 2.55 \times 10^{-2}$
$= \underline{2.6} \times 10^{-2}$

the number $\underline{6.8}$ has only two significant digits; the quotient 2.55 was, therefore, rounded to $\underline{2.6}$ with two significant digits

$\left(\text{Note: } \dfrac{10^{-5}}{10^{-3}} = 10^{-5} \times 10^3 = 10^{-2}\right)$

(b) $\dfrac{.00003}{1,200,000} = \dfrac{\underline{3} \times 10^{-5}}{1.2 \times 10^6}$
$= \dfrac{\underline{3}}{1.2} \times \dfrac{10^{-5}}{10^6}$
$= 2.5 \times 10^{-11}$
$= \underline{3} \times 10^{-11}$

the number $\underline{3}$ has only one significant digit; the quotient .25 was rounded to $\underline{3}$ with one significant digit

$3 \div 1.2 = 2.5$

$\left(\text{Note: } \dfrac{10^{-5}}{10^6} = 10^{-5} \times 10^{-6} = 10^{-11}\right)$

(c) $37,000,000 \times 5,200,000$
$= 3.7 \times 10^7 \times 5.2 \times 10^6$
$= 19 \times 10^{13}$

$\underline{3.7} \times \underline{5.2} = 19.24$ rounds to $\underline{19}$
$10^7 \times 10^6 = 10^{13}$

(d) $.0007 \times 83{,}000 = 7 \times 10^{-4} \times 8.3 \times 10^{4}$ $\underline{7} \times 8.3 = 58.1$
$= 60 \times 10^{0}$ rounds to $\underline{60}$
$= 60$ recall: $10^{0} = 1$

(e) $\dfrac{.00084}{.0032} = \dfrac{8.4 \times 10^{-4}}{3.2 \times 10^{-3}}$ $\underline{8.4} \div \underline{3.2} = 2.625$ rounds to $\underline{2.6}$
$= 2.6 \times 10^{-1}$ $\dfrac{10^{-4}}{10^{-3}} = 10^{-4} \times 10^{3} = 10^{-1}$

(f) $\dfrac{.005}{28{,}000} = \dfrac{5 \times 10^{-3}}{2.8 \times 10^{4}}$ $\underline{5} \div 2.8 = 1.785$ rounds to $\underline{2}$
$= 2 \times 10^{-7}$ $\dfrac{10^{-3}}{10^{4}} = 10^{-3} \times 10^{-4} = 10^{-7}$

Astronomy

(g) The distance light travels in 1 year is called a **light-year**. This distance is approximately 5.9×10^{12} miles. The Andromeda galaxy is about 2×10^{6} light-years away from the earth. This means that it takes light from Andromeda $2 \times 10^{6} = 2{,}000{,}000$ years to reach the earth. How many miles is the Andromeda galaxy from the earth?

Solution: Using 5.9×10^{12} mi $= 1$ light-year,

$$2 \times 10^{6} \text{ light-years} = 2 \times 10^{6} \text{ light-years} \times \dfrac{5.9 \times 10^{12} \text{ mi}}{1 \text{ light-year}}$$

$$= \underline{2} \times 10^{6} \times 5.9 \times 10^{12} \text{ mi}$$

$$= 11.8 \times 10^{18} \text{ mi}$$

$$= \underline{10} \times 10^{18} \text{ mi}$$

$$= 10^{19} \text{ mi}$$

Since $\underline{2}$ has only one significant digit, the product 11.8 was rounded to one significant digit.

Astronomy

(h) Stars that are of a large enough size eventually collpase and become neutron stars. Neutron stars have a radius of about 6 miles and a density of 3×10^{14} grams per cubic centimeter. If a white dwarf has a density of 1×10^{6} grams per cubic centimeter, how many times greater is the density of a neutron star?

Solution: $\dfrac{3 \times 10^{14}}{1 \times 10^{6}} = 3 \times 10^{8}$ times as dense

Astronomy

(i) Assuming the density of a white dwarf is 1.0×10^{6} grams per cubic centimeter, how many tons would 1 cubic centimeter weigh?

Solution: Using 1 kg = 1,000 g = 10^3 g and 1 ton = 2,000 lb = 2×10^3 lb, we have

$$1 \text{ cm}^3 = 1 \text{ cm}^3 \times \frac{1.0 \times 10^6 \text{ g}}{1 \text{ cm}^3} \times \frac{1 \text{ kg}}{10^3 \text{ g}} \times \frac{2.20 \text{ lb}}{1 \text{ kg}} \times \frac{1 \text{ ton}}{2 \times 10^3 \text{ lb}}$$

$$= \frac{1.0 \times 10^6 \times 2.20 \text{ tons}}{10^3 \times 2 \times 10^3}$$

$$= \frac{2.2 \times 10^6}{2 \times 10^6} \quad \text{since the digit 2 is exact, our answer has the same number of significant digits as 2.2 has}$$

$$= 1.1 \text{ tons}$$

Astronomy

(j) How bright a star appears to us depends on how bright the star is and how far away the star is. If two stars are equally bright, the closer star appears to be brighter. The number of times brighter can be computed using the formula

$$N = \left(\frac{\text{distance to farther star}}{\text{distance to closer star}}\right)^2$$

As an example, if two stars of equal brightness are 8 light-years and 4 light-years away, the closer star will appear to be 4 times as bright. This is shown below.

$$N = \left(\frac{8 \text{ light-years}}{4 \text{ light-years}}\right)^2 = (2)^2 = 4$$

The sun and Alpha Centauri, the star closest to the sun, are of about equal brightness. If the distances of Alpha Centauri and the sun from the earth are 2.4×10^{13} miles and 9.3×10^7 miles, respectively, how many times as bright does the sun appear to us?

Solution:

$$N = \left(\frac{2.4 \times 10^{13}}{9.3 \times 10^7}\right)^2 = (.26 \times 10^6)^2 \qquad 2.4 \div 9.4 = .26$$
$$\qquad\qquad\qquad\qquad\qquad\qquad\qquad\qquad 10^{13} \div 10^7 = 10^6$$

$$= .26 \times 10^6 \times .26 \times 10^6$$

$$= .068 \times 10^{12}$$

$$= .068 \times 10^2 \times 10^{10}$$

$$= 6.8 \times 10^{10}$$

$$= 68,000,000,000 \text{ times as bright}$$

Sec. 12-3 Scientific Notation

Complete the problems below. Completed problems appear on page 466.

Sample Set

(a) $5{,}700{,}000 = 57 \times 10^{\square} = 5.7 \times 10^{\square} = .57 \times 10^{\square}$

(b) $.0000372 = .372 \times 10^{\square} = 3.72 \times 10^{\square} = 37.2 \times 10^{\square}$

Write each number using standard notation.

(c) $.75 \times 10^8 = $ _____

(d) $7.5 \times 10^8 = $ _____

(e) $75 \times 10^8 = $ _____

(f) $.75 \times 10^{-8} = $ _____

(g) $7.5 \times 10^{-8} = $ _____

(h) $75 \times 10^{-8} = $ _____

Compute each product. Round the answer to the appropriate number of significant digits.

(i) $(8.3 \times 10^7)(3.43 \times 10^8) = 28.469 \times 10^{\square} = \square \times 10^{15}$
$= \square \times 10^{16}$

(j) $(5.6 \times 10^{-5})(\underline{7} \times 10^{-4}) = 39.2 \times 10^{\square} = \square \times 10^{-9}$
$= \square \times 10^{-8}$

(k) $(6.7 \times 10^{-8})(5.6 \times 10^{17}) = 37.52 \times 10^{\square} = \square \times 10^{9}$
$= \square \times 10^{10}$

Compute each quotient. Round the answer to the appropriate number of significant digits.

(l) $\dfrac{8 \times 10^{12}}{1.9 \times 10^{-5}} = 4.21 \times 10^{\square} = \square \times 10^{17}$

(m) $\dfrac{4.2 \times 10^{-5}}{9.3 \times 10^{-6}} = .451 \times 10^{\square} = \square \times 10^{1} = \square$

(n) $\dfrac{3.4 \times 10^8}{7 \times 10^{-5}} = .48 \times 10^{\square} = \square \times 10^{13} = \square \times 10^{1} \times 10^{12}$
$= \square \times 10^{12}$

Completed Problems

(a) $5{,}700{,}000 = 57 \times 10^{\boxed{5}} = 5.7 \times 10^{\boxed{6}} = .57 \times 10^{\boxed{7}}$

(b) $.0000372 = .372 \times 10^{\boxed{-4}} = 3.72 \times 10^{\boxed{-5}} = 37.2 \times 10^{\boxed{-6}}$

(c) $.75 \times 10^8 = \boxed{75{,}000{,}000}$

(d) $7.5 \times 10^8 = \boxed{750{,}000{,}000}$

(e) $75 \times 10^8 = \boxed{7{,}500{,}000{,}000}$

(f) $.75 \times 10^{-8} = \boxed{.0000000075}$

(g) $7.5 \times 10^{-8} = \boxed{.000000075}$

(h) $75 \times 10^{-8} = \boxed{.00000075}$

(i) $(8.3 \times 10^7)(3.43 \times 10^8) = 28.469 \times 10^{\boxed{15}} = \boxed{28} \times 10^{15}$
$= \boxed{2.8} \times 10^{16}$

(j) $(5.6 \times 10^{-5})(\underline{7} \times 10^{-4}) = 39.2 \times 10^{\boxed{-9}} = \boxed{40} \times 10^{-9}$
$= \boxed{4} \times 10^{-8}$

(k) $(6.7 \times 10^{-8})(5.6 \times 10^{17}) = 37.52 \times 10^{\boxed{9}} = \boxed{38} \times 10^9$
$= \boxed{3.8} \times 10^{10}$

(l) $\dfrac{8 \times 10^{12}}{1.9 \times 10^{-5}} = 4.21 \times 10^{\boxed{17}} = \boxed{4} \times 10^{17}$

(m) $\dfrac{4.2 \times 10^{-5}}{9.3 \times 10^{-6}} = .451 \times 10^{\boxed{1}} = \boxed{.45} \times 10^1 = \boxed{4.5}$

(n) $\dfrac{3.4 \times 10^8}{7 \times 10^{-5}} = .48 \times 10^{\boxed{13}} = \boxed{.5} \times 10^{13} = \boxed{.5} \times 10^1 \times 10^{12}$
$= \boxed{5} \times 10^{12}$

Name: _____

Class: _____

Exercises for Sec. 12-3 Compute each answer using scientific notation. Round your answer to the appropriate number of significant digits.

1. $850{,}000 \times 4{,}300{,}000$

2. $7{,}200 \times 58{,}000{,}000$

3. $59{,}000 \times 700{,}000{,}000$

4. $.00091 \times 6{,}200{,}000$

5. $.00004 \times 85{,}000{,}000$

6. $.000009 \times 7{,}000{,}000{,}000$

7. $.0005 \times .0000064$

8. $.000072 \times .00318$

9. $.0000081 \times .00000092$

10. $\dfrac{36{,}000{,}000}{540{,}000}$

11. $\dfrac{280{,}000}{96{,}000{,}000}$

12. $\dfrac{8{,}000{,}000{,}000}{73{,}000{,}000}$

13. $\dfrac{.00024}{490{,}000}$

14. $\dfrac{.000004}{5{,}000{,}000}$

15. $\dfrac{8{,}900{,}000}{.00062}$

16. $\dfrac{8{,}000{,}000{,}000}{.0000053}$

17. $\dfrac{.000084}{.0007}$

18. $\dfrac{.0000056}{.0019}$

468 Exponents and Roots

19. $\dfrac{.00047}{.0000092}$

20. $\dfrac{.0005}{.0000078}$

Think-big problems

21. The red star Betelgeuse in the constellation Orion is 520 light-years away from the earth. How many miles is Betelgeuse from the earth? (See Example g.)

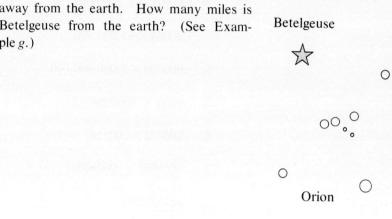

22. The Whirlpool galaxy is 2×10^7 light-years away from the earth. How many miles is the Whirlpool galaxy from the earth? (See Example g.)

23. The most distant objects observed in the universe are quasars. A quasar that is 1.1×10^{10} light-years from the earth is how many times farther away than the Andromeda galaxy? (The Andromeda galaxy, which is 2×10^6 light-years away, might be considered a close neighbor by comparison.)

24. The diameter of the Milky Way galaxy is approximately 5.9×10^{17} miles. Using 5.9×10^{12} miles as the distance light travels in 1 year, how long would it take light to travel from one side of our galaxy to the other?

25. The largest stars in the universe are believed to eventually collapse into bodies called black holes. Their gravitational pull becomes so great that nearby stars are drawn to them and disappear. Even light cannot escape their gravitational pull. It is believed that the radius of a black hole is about 2 miles, and that their density is 8×10^{15} grams per cubic centimeter. How many times greater is the density of a black hole than the density of a white dwarf? (See Example h.)

26. How many times greater is the density of a black hole than the density of a neutron star? (See Exercise 25 and Example h.)

27. How many tons would 1 cubic centimeter of a black hole weigh? (See Exercise 25 and Example i.)

28. The mass of the earth is estimated to be 6.588×10^{13} tons. Find the mass in kilograms. (Use 1 ton = 2×10^3 pounds, and 2.2 pounds = 1 kilogram.)
    (*Answer:* 6.0×10^{16} kg)

29. Find the mass in kilograms of 1 cubic meter of a black hole. (Use density = 8×10^{15} grams per cubic centimeter; 1 cubic meter = 10^6 cubic centimeters; and 1 kilogram = 10^3 grams.)
(*Answer:* 8×10^{18} kg)

30. How many times greater is the mass of 1 cubic meter of a black hole than the entire mass of the earth? (See the answers to Exercises 28 and 29. Round your answer to one significant digit.)

31. Assume that an average star in the Andromeda galaxy is as bright as the star Alpha Centauri. Alpha Centauri is 4 light-years away from the earth, and the Andromeda galaxy is 2×10^6 light-years away. How many times brighter would Alpha Centauri appear to us than an average star in the Andromeda galaxy? (See Example *j*.)
(*Answer:* $.25 \times 10^{12}$ or $.3 \times 10^{12}$ times as bright)

32. It is estimated that there are 10^{11} stars in the Andromeda galaxy. The entire galaxy would appear that many times brighter than the average star in the galaxy. How many times brighter is Alpha Centauri than the entire Andromeda galaxy? (Divide 10^{11} by the answer to Exercise 31.)
(*Answer:* 2.5 times as bright)

Think-small problems

33. A hydrogen atom has one proton and one electron. The mass of a proton is 1.67×10^{-24} gram, and the mass of an electron is $.91 \times 10^{-27}$ gram. A proton is how many times heavier than an electron?

34. The radius of a hydrogen atom is 5.3×10^{-9} centimeter. If one could line up hydrogen atoms side by side, how many atoms would it take to cover a length of 1 inch?

35. Using 1.67×10^{-24} gram for the weight of a hydrogen atom, how many atoms would be in $\underline{1.00}$ gram of hydrogen (to three significant digits)?

Section 12-4 Square and Cube Roots of Whole Numbers

Square Roots Every number can be written as a product of two equal factors. To find a **square root** of a number is to find **one** of these factors. For example,

3 is a square root of 9	since	$(3)(3) = 9$
5 is a square root of 25	since	$(5)(5) = 25$
-5 is also a square root of 25	since	$(-5)(-5) = 25$

The **positive square root** of a number is indicated using the **radical sign**, $\sqrt{}$.

$$\sqrt{9} = 3 \quad \text{not} \; -3$$

A number whose square root is a whole number is called a **perfect square**. The perfect squares less than or equal to 100 are 1, 4, 9, 16, 25, 36, 49, 64, 81, and 100.

Square roots of numbers that are not perfect squares can be approximated. Some examples, approximated to the nearest 1,000th, are given below.

$$\sqrt{2} = 1.414 \quad \sqrt{5} = 2.236$$
$$\sqrt{3} = 1.732 \quad \sqrt{7} = 2.646$$

A method for approximating square roots is given at the end of this section.

Have you asked yourself: "What is the square root of a negative number such as -4?" The square root cannot be -2 since $(-2)(-2) \neq -4$. It cannot be 2, since $(2)(2) \neq -4$. In more advanced texts, you will find the number system extended to include what are called **imaginary** numbers. In this larger system, every number, including the negative numbers, will have two square roots.

The property of square roots that will be used in this section is given below.

Multiplication Rule

For positive numbers a and b,
$$\sqrt{ab} = \sqrt{a}\sqrt{b}$$

In words, the square root of a product of positive factors is the product of the square roots of the factors. The truth of this statement can be demonstrated using the product $(4)(9)$.

$$\sqrt{36} = 6 \quad \text{since } (6)(6) = 36$$

or
$$\sqrt{36} = \sqrt{(4)(9)}$$
$$= \sqrt{4}\sqrt{9}$$
$$= (2)(3)$$
$$= 6$$

474 Exponents and Roots

The above example demonstrates more than just the multiplication rule. It shows us that **the product of perfect squares is itself a perfect square.**

Sample Problem 12-2

Solution:

Estimate $\sqrt{12}$ using $\sqrt{3} = 1.732$.

Factor 12 so that one factor is a perfect square; the perfect square is 4:

$$12 = \sqrt{4}\sqrt{3}$$
$$= (2)(1.732)$$
$$= 3.464 \quad \text{answer}$$

Is 3.464 a good approximation? Would 3.463, or 3.465, have been better?

$$(3.463)(3.463) = 11.992369$$
$$(3.464)(3.464) = 11.999296$$
$$(3.465)(3.465) = 12.006225$$

Note that 11.999296 is closer to 12 than is either 11.992369 or 12.006225. To the nearest 1,000th, 3.464 is, therefore, the best approximation.

It would have been possible to estimate $\sqrt{12}$ using $\sqrt{2} = 1.414$ and $\sqrt{3} = 1.732$.

$\sqrt{12} = \sqrt{(2)(2)(3)}$
$\phantom{\sqrt{12}} = \sqrt{2}\sqrt{2}\sqrt{3}$
$\phantom{\sqrt{12}} = (1.414)(1.414)(1.732)$
$\phantom{\sqrt{12}} = (1.999396)(1.732)$
$\phantom{\sqrt{12}} = 3.462953872$
$\phantom{\sqrt{12}} = \underline{3.463}$

since each root is an approximation with four significant digits, the final product was rounded to four significant digits

Note two things.

1. The estimation 3.463 of $\sqrt{12}$ is not as accurate as that found in the sample problem. This is due to the fact that three approximations were used instead of one.
2. More work is required when one of the factors is not a perfect square.

This demonstrates the advantage of having a perfect square as one of the factors.

Sample Problem 12-3

Solution:

Estimate $\sqrt{48}$ using $\sqrt{3} = 1.732$.

Factor 48 so that one factor is a perfect square; the perfect square is 16:

$$48 = \sqrt{16}\sqrt{3}$$
$$= (4)(1.732)$$
$$= 6.928$$

Sec. 12-4 Square and Cube Roots of Whole Numbers

> A **common error** is to compute a second root of a factor. For example,
>
> $$48 = \sqrt{16}\,\sqrt{3} \qquad \text{thinking that } \sqrt{4} = 2,$$
> $$= (4)(1.732) \qquad \text{students often write}$$
> $$= (2)(1.732) \qquad \leftarrow \text{this; it is \textbf{wrong}!}$$
> $$= 3.464$$
>
> When roots were written in the second line above, no more roots are to be computed.

It is often difficult to see the largest perfect square that is a factor of a number. If this is the case, smaller perfect squares may be used. This is demonstrated in the next sample problem.

Sample Problem 12-4 Estimate $\sqrt{180}$ using $\sqrt{5} = 2.236$.

Solution:

The smallest perfect square, other than 1, is 4; try dividing by 4:

$$\begin{array}{r}4\,)\,180\\ \hline 45\end{array}$$

Since 45 is divisible by the perfect square 9, divide by 9:

$$\begin{array}{r}4\,)\,180\\ 9\,)\,45\\ \hline 5\end{array}$$

$\sqrt{180}$ can now be estimated:

2.236 has four significant digits, so the product was rounded to four significant digits:

$$\sqrt{180} = \sqrt{4}\,\sqrt{9}\,\sqrt{5}$$
$$= (2)(3)(2.236)$$
$$= (6)\,(2.236)$$
$$= 13.416$$
$$= \underline{13.42}$$

In the problem above, the factor 6 is exactly equal to $\sqrt{4}\,\sqrt{9}$; not an approximation. It was not considered when counting significant digits.

Cube Roots Every number can be written as a product of three equal factors. To find a **cube root** is to find one of those factors. The cube root of a number is indicated using the radical symbol, $\sqrt[3]{}$.

$$\sqrt[3]{8} = 2 \qquad \text{since} \qquad (2)(2)(2) = 8$$
$$\sqrt[3]{-8} = -2 \qquad \text{since} \qquad (-2)(-2)(-2) = -8$$
$$\sqrt[3]{27} = 3 \qquad \text{since} \qquad (3)(3)(3) = 27$$
$$\sqrt[3]{-27} = -3 \qquad \text{since} \qquad (-3)(-3)(-3) = -27$$

Numbers whose cube roots are whole numbers are called **perfect cubes**. Examples of perfect cubes are 1, 8, and 27. As in the case of square roots, cube roots of numbers that are not perfect cubes can be approximated. Some examples are

$$\sqrt[3]{2} = 1.260 \qquad \sqrt[3]{5} = 1.710$$
$$\sqrt[3]{3} = 1.442 \qquad \sqrt[3]{7} = 1.913$$

476   Exponents and Roots

Complete the problems below. Completed problems appear on page 478.

Sample Set

In each problem, the number can be written as a product of a perfect square and one of the numbers 2, 3, 5, and 7. Using $\sqrt{2} = 1.414$, $\sqrt{3} = 1.732$, $\sqrt{5} = 2.236$, or $\sqrt{7} = 2.646$, estimate the root. Round the answer to the appropriate number of significant digits.

(a) $\sqrt{18} = \sqrt{}\ \sqrt{2}$

$= ()(1.414)$

$= \boxed{}$

(b) $\sqrt{147} = \sqrt{49}\ \sqrt{}$

$= (7)()$

$= \boxed{}$

$= \boxed{}$

(c) $\sqrt{392} = \sqrt{}\ \sqrt{}\ \sqrt{2}$

$= ()()()$

$= \boxed{}$

$= \boxed{}$

$\begin{array}{r} 4\,)\,392 \\ 49\,)\,98 \\ \hline 2 \end{array}$

In each problem, the number can be written as a product of a perfect cube and one of the numbers 2, 3, 4, 5, 6, and 7. Using $\sqrt[3]{2} = 1.260$, $\sqrt[3]{3} = 1.442$, $\sqrt[3]{4} = 1.587$, $\sqrt[3]{5} = 1.710$, $\sqrt[3]{6} = 1.817$, or $\sqrt[3]{7} = 1.913$, estimate the root.

(d) $\sqrt[3]{24} = \sqrt[3]{}\ \sqrt[3]{3}$

$= ()(1.442)$

$= \boxed{}$

(e) $\sqrt[3]{108} = \sqrt[3]{27}\ \sqrt[3]{}$

$= (3)()$

$= \boxed{}$

The property of cube roots that will be used in this section is given below.

Multiplication Rule

$\sqrt[3]{ab} = \sqrt[3]{a}\ \sqrt[3]{b}$

The truth of this rule can be demonstrated using the product $(8)(27)$.

$\sqrt[3]{216} = 6 \qquad \text{since } (6)(6)(6) = 216$

or $\quad \sqrt[3]{216} = \sqrt[3]{(8)(27)}$

$= \sqrt[3]{8}\ \sqrt[3]{27}$

$= (2)\ (3)$

$= 6$

Sample Problem 12-5 Estimate $\sqrt[3]{54}$ using $\sqrt[3]{2} = 1.260$.

Solution: Factor 54 so that one factor is a perfect cube:
$$54 = \sqrt[3]{27} \cdot \sqrt[3]{2}$$
$$= (3)(1.260)$$
$$= 3.780$$

Approximation of Roots

The method that will be given here for approximating the square root of 2 is not the most efficient one known. However, it will lead us to an important property of the number system.

As a first estimate of $\sqrt{2}$, try 1.

$$1^2 = 1 < 2$$

1 is too small, so try 2.

$$2^2 = 4 > 2$$

2 is too large an estimate. The desired number must be between 1 and 2.

As a third estimate, try the number halfway between 1 and 2. This number is called the **midpoint** between 1 and 2, and can be found by dividing the sum of 1 and 2 by 2.

$$\text{Midpoint} = \frac{1+2}{2} = \frac{3}{2} = 1.5$$

```
  1           1.5           2
  •            •            •
first:       third:       second:
estimate:   estimate:    estimate:
too small   too big      too big
```

$$(1.5)^2 = 2.25 > 2$$

1.5 is too large, so the desired number is between 1 and 1.5. As a fourth estimate, try the midpoint between 1 and 1.5.

$$\text{Midpoint} = \frac{1+1.5}{2} = \frac{2.5}{2} = 1.25$$

```
  1     1.25    1.5              2
  •      •      •                •
first:  fourth: third:         second:
too     too     too            too
small   small   big            big
```

$$(1.25)^2 = 1.5625 < 2$$

1.25 is too small, so the desired number is between 1.25 and 1.5. As a fifth estimate, try the midpoint between 1.25 and 1.5.

$$\text{Midpoint} = \frac{1.25 + 1.5}{2} = 1.375$$

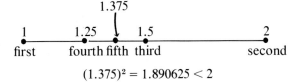

$$(1.375)^2 = 1.890625 < 2$$

1.375 is too small, so the desired number is between 1.375 and 1.5. The sixth estimate is the midpoint between 1.375 and 1.5.

$$\text{Midpoint} = \frac{1.375 + 1.5}{2} = 1.4375$$

$(1.4375)^2 = 2.0664062$, very close to 2.

Note that numbers are being found whose squares are getting closer to 2. It appears that these numbers are narrowing in on a point which will represent the true square root of 2. Does such a point really exist? In order to guarantee that such a point does exist, mathematicians have set forth a law called the **completeness property.** The completeness property essentially states that the number line is solid; that is, complete and without holes. This property cannot be proved. It is simply accepted so that one can state with assurance that an approximation process, such as the one used above, leads to an answer.

If we accept the completeness property, we can conclude that the approximations given above do approach a number which will be the square root of 2.

Completed Problems

(a) $\sqrt{18} = \sqrt{9}\sqrt{2}$
$= (3)(1.414)$
$= \boxed{4.242}$

(b) $\sqrt{147} = \sqrt{49}\sqrt{3}$
$= (7)(1.732)$
$= \boxed{12.124}$
$= \boxed{12.12}$

(c) $\sqrt{392} = \sqrt{4}\sqrt{49}\sqrt{2}$
$= (2)(7)(1.414)$
$= \boxed{19.796}$
$= \boxed{19.80}$

(d) $\sqrt[3]{24} = \sqrt[3]{8}\sqrt[3]{3}$
$= (2)(1.442)$
$= \boxed{2.884}$

(e) $\sqrt[3]{108} = \sqrt[3]{27}\sqrt[3]{4}$
$= (3)(1.587)$
$= \boxed{4.761}$

Name: _____

Class: _____

Exercises for Sec. 12-4 Estimate the square roots using $\sqrt{2} = 1.414$, $\sqrt{3} = 1.732$, $\sqrt{5} = 2.236$, or $\sqrt{7} = 2.646$. Round each answer to the appropriate number of significant digits.

1. $\sqrt{8} =$ 2. $\sqrt{27} =$ 3. $\sqrt{45} =$

4. $\sqrt{20} =$ 5. $\sqrt{32} =$ 6. $\sqrt{80} =$

7. $\sqrt{72} =$ 8. $\sqrt{108} =$ 9. $\sqrt{28} =$

10. $\sqrt{63} =$ 11. $\sqrt{50} =$ 12. $\sqrt{75} =$

13. $\sqrt{98} =$ 14. $\sqrt{196} =$ 15. $\sqrt{144} =$

16. $\sqrt{192} =$ 17. $\sqrt{125} =$ 18. $\sqrt{112} =$

19. $\sqrt{288} =$ 20. $\sqrt{720} =$

Estimate the cube roots using $\sqrt[3]{2} = 1.260$, $\sqrt[3]{3} = 1.442$, $\sqrt[3]{4} = 1.587$, $\sqrt[3]{5} = 1.710$, $\sqrt[3]{6} = 1.817$, or $\sqrt[3]{7} = 1.913$. Round the answer in Exercises 39 and 40 to the appropriate number of significant digits.

21. $\sqrt[3]{16} =$ 22. $\sqrt[3]{81} =$ 23. $\sqrt[3]{135} =$

24. $\sqrt[3]{162} =$ 25. $\sqrt[3]{32} =$ 26. $\sqrt[3]{40} =$

27. $\sqrt[3]{189} =$ 28. $\sqrt[3]{48} =$ 29. $\sqrt[3]{56} =$

30. $\sqrt[3]{128} =$ 31. $\sqrt[3]{192} =$ 32. $\sqrt[3]{320} =$

33. $\sqrt[3]{256} =$ 34. $\sqrt[3]{512} =$ 35. $\sqrt[3]{432} =$

36. $\sqrt[3]{448} =$ 37. $\sqrt[3]{729} =$ 38. $\sqrt[3]{1080} =$

39. $\sqrt[3]{1458} =$ 40. $\sqrt[3]{1512} =$

Section 12-5 Square and Cube Roots of Fractions

Square Roots The property of square roots that will be used in this section is given below.

Division Rule

$$\sqrt{\frac{a}{b}} = \frac{\sqrt{a}}{\sqrt{b}}$$

Examples

(a) Compute $\sqrt{\frac{4}{9}}$.

$$\sqrt{\frac{4}{9}} = \frac{\sqrt{4}}{\sqrt{9}} = \frac{2}{3} \quad \text{Check: } \left(\frac{2}{3}\right)\left(\frac{2}{3}\right) = \frac{4}{9}$$

(b) Estimate $\sqrt{\frac{3}{4}}$.

$$\sqrt{\frac{3}{4}} = \frac{\sqrt{3}}{\sqrt{4}} = \frac{1.732}{2} \quad \text{using } \sqrt{3} = 1.732$$

$$= .866$$

Is .866 a good estimate?
Check: $(.866)^2 = .749956$, which is very close to $.75 = \frac{3}{4}$.

(c) Estimate $\sqrt{\frac{1}{3}}$. In this example, the denominator is not a perfect square.

Method 1 (the hard way)

$$\sqrt{\frac{1}{3}} = \frac{\sqrt{1}}{\sqrt{3}} = \frac{1}{1.732} \quad \text{dividing 1 by 1.732 and rounding the quotient to four significant digits}$$

$$= .57736\cdots$$

$$= \underline{.5774}$$

Method 2 (the easy way)
The problem would be easier if the denominator were a whole number. This can be accomplished by multiplying the fraction by $\frac{\sqrt{3}}{\sqrt{3}} = 1$.

$$\sqrt{\frac{1}{3}} = \frac{\sqrt{1}}{\sqrt{3}} = \frac{1}{\sqrt{3}} \times \frac{\sqrt{3}}{\sqrt{3}} \quad \begin{array}{l}\sqrt{3}\,\sqrt{3} = 3 \text{ by the definition of } \sqrt{3}, \text{ or}\\ \sqrt{3}\,\sqrt{3} = \sqrt{(3)(3)} = \sqrt{9} = 3\end{array}$$

$$= \frac{\sqrt{3}}{3}$$

$$= \frac{1.732}{3}$$

$$= .57733$$

$$= \underline{.5773}$$

This method of making the denominator a whole number is called **rationalizing the denominator.** The difference in the answers is caused by the different approaches used to work the problem. The rule for determining the number of significant digits in an answer may leave the last digit in error.

(d) Estimate $\sqrt{\frac{1}{8}}$.

$$\sqrt{\frac{1}{8}} = \frac{\sqrt{1}}{\sqrt{8}} = \frac{1}{\sqrt{8}}\frac{\sqrt{2}}{\sqrt{2}}$$ 16 is the smallest perfect square divisible by 8. To obtain $\sqrt{16}$, multiply $\sqrt{8}$ by $\sqrt{2}$

$$= \frac{\sqrt{2}}{\sqrt{16}}$$

$$= \frac{1.414}{4} \quad \text{using } \sqrt{2} = 1.414$$

$$= .3535$$

(e) Estimate $\sqrt{\frac{5}{12}}$.

$$\sqrt{\frac{5}{12}} = \frac{\sqrt{5}}{\sqrt{12}} = \frac{\sqrt{5}}{\sqrt{12}}\frac{\sqrt{3}}{\sqrt{3}} \quad \text{the smallest perfect square divisible by 12 is 36}$$

$$= \frac{\sqrt{5}\sqrt{3}}{\sqrt{36}}$$

$$= \frac{(2.236)(1.732)}{6} \quad \text{using } \sqrt{5} = 2.236 \text{ and } \sqrt{3} = 1.732$$

$$= \frac{3.873}{6}$$

$$= .6455$$

Cube Roots The division rule for cube roots is given below.

Division Rule $\sqrt[3]{\frac{a}{b}} = \frac{\sqrt[3]{a}}{\sqrt[3]{b}}$

Examples

(a) Compute $\sqrt[3]{\frac{27}{8}}$.

$$\sqrt[3]{\frac{27}{8}} = \frac{\sqrt[3]{27}}{\sqrt[3]{8}} = \frac{3}{2} \quad \text{Check: } \left(\frac{3}{2}\right)\left(\frac{3}{2}\right)\left(\frac{3}{2}\right) = \frac{27}{8}$$

(b) Estimate $\sqrt[3]{\frac{5}{8}}$.

$$\sqrt[3]{\frac{5}{8}} = \frac{\sqrt[3]{5}}{\sqrt[3]{8}} = \frac{1.710}{2} = .855$$

(c) Estimate $\sqrt[3]{\dfrac{2}{9}}$.

$$\sqrt[3]{\dfrac{2}{9}} = \dfrac{\sqrt[3]{2}}{\sqrt[3]{9}} = \dfrac{\sqrt[3]{2}}{\sqrt[3]{9}} \dfrac{\sqrt[3]{3}}{\sqrt[3]{3}} \qquad \text{27 is the smallest perfect cube divisible by 9; to obtain } \sqrt[3]{27}, \text{ multiply } \sqrt[3]{9} \text{ by } \sqrt[3]{3}$$

$$= \dfrac{\sqrt[3]{2}\,\sqrt[3]{3}}{\sqrt[3]{27}}$$

$$= \dfrac{(1.260)(1.442)}{3} \qquad \text{using } \sqrt[3]{2} = 1.260 \text{ and } \sqrt[3]{3} = 1.442$$

$$= \dfrac{1.817}{3}$$

$$= .6057$$

Complete the problems below. Completed answers appear on page 485.

Sample Set

Estimate the roots using $\sqrt{2} = 1.414$, $\sqrt{3} = 1.732$, $\sqrt{5} = 2.236$, or $\sqrt{7} = 2.646$. Round the answer to the appropriate number of significant digits.

(a) $\sqrt{\dfrac{5}{16}} = \dfrac{\sqrt{5}}{\sqrt{16}} = \dfrac{\boxed{}}{\boxed{}} = \boxed{}$

(b) $\sqrt{\dfrac{1}{2}} = \dfrac{\sqrt{1}}{\sqrt{2}} = \dfrac{1}{\sqrt{2}}\dfrac{\sqrt{2}}{\sqrt{2}} = \dfrac{\sqrt{2}}{\sqrt{4}} = \dfrac{\boxed{}}{\boxed{}} = \boxed{}$

(c) $\sqrt{\dfrac{1}{27}} = \dfrac{\sqrt{1}}{\sqrt{27}} = \dfrac{1}{\sqrt{27}}\dfrac{\sqrt{3}}{\sqrt{3}} = \dfrac{\sqrt{}}{\sqrt{}} = \dfrac{\boxed{}}{\boxed{}} = \boxed{}$

(d) $\sqrt{\dfrac{7}{8}} = \dfrac{\sqrt{7}}{\sqrt{8}} = \dfrac{\sqrt{7}}{\sqrt{8}}\dfrac{\sqrt{}}{\sqrt{}} = \dfrac{\sqrt{7}\,\sqrt{}}{\sqrt{}} = \dfrac{()()}{\boxed{}}$

$= \dfrac{\boxed{}}{\boxed{}} = \dfrac{\boxed{}}{\boxed{}} = \boxed{}$

Estimate the roots using $\sqrt[3]{2} = 1.260$, $\sqrt[3]{3} = 1.442$, $\sqrt[3]{4} = 1.587$, $\sqrt[3]{5} = 1.710$, $\sqrt[3]{6} = 1.817$, or $\sqrt[3]{7} = 1.913$. Round each answer to the appropriate number of significant digits.

(e) $\sqrt[3]{\dfrac{7}{27}} = \dfrac{\sqrt[3]{7}}{\sqrt[3]{27}} = \dfrac{\boxed{}}{\boxed{}} = \boxed{}$

(f) $\sqrt[3]{\dfrac{1}{4}} = \dfrac{\sqrt[3]{1}}{\sqrt[3]{4}} = \dfrac{1}{\sqrt[3]{4}} \dfrac{\sqrt[3]{}}{\sqrt[3]{}} = \dfrac{\sqrt[3]{}}{\sqrt[3]{}} = \dfrac{\boxed{}}{\boxed{}} = \boxed{}$

(g) $\sqrt[3]{\dfrac{5}{9}} = \dfrac{\sqrt[3]{5}}{\sqrt[3]{9}} = \dfrac{\sqrt[3]{5}}{\sqrt[3]{9}} \dfrac{\sqrt[3]{}}{\sqrt[3]{}} = \dfrac{\sqrt[3]{5} \; \sqrt[3]{}}{\sqrt[3]{}}$

$= \dfrac{()()}{\boxed{}} = \dfrac{\boxed{}}{\boxed{}}$

$= \dfrac{\boxed{}}{\boxed{}} = \boxed{}$

Completed Problems

(a) $\sqrt{\dfrac{5}{16}} = \dfrac{\sqrt{5}}{\sqrt{16}} = \dfrac{2.236}{4} = .5990$

(b) $\sqrt{\dfrac{1}{2}} = \dfrac{\sqrt{1}}{\sqrt{2}} = \dfrac{1}{\sqrt{2}} \cdot \dfrac{\sqrt{2}}{\sqrt{2}} = \dfrac{\sqrt{2}}{\sqrt{4}} = \dfrac{1.414}{2} = .7070$

(c) $\sqrt{\dfrac{1}{27}} = \dfrac{\sqrt{1}}{\sqrt{27}} = \dfrac{1}{\sqrt{27}} \cdot \dfrac{\sqrt{3}}{\sqrt{3}} = \dfrac{\sqrt{3}}{\sqrt{81}} = \dfrac{1.732}{9} = .1924$

(d) $\sqrt{\dfrac{7}{8}} = \dfrac{\sqrt{7}}{\sqrt{8}} = \dfrac{\sqrt{7}}{\sqrt{8}} \cdot \dfrac{\sqrt{2}}{\sqrt{2}} = \dfrac{\sqrt{7}\,\sqrt{2}}{\sqrt{16}} = \dfrac{(2.646)(1.414)}{4}$

$= \dfrac{3.7414\,444}{4} = \dfrac{3.741}{4} = .9353$

(e) $\sqrt[3]{\dfrac{7}{27}} = \dfrac{\sqrt[3]{7}}{\sqrt[3]{27}} = \dfrac{1.913}{3} = .6377$

(f) $\sqrt[3]{\dfrac{1}{4}} = \dfrac{\sqrt[3]{1}}{\sqrt[3]{4}} = \dfrac{1}{\sqrt[3]{4}} \cdot \dfrac{\sqrt[3]{2}}{\sqrt[3]{2}} = \dfrac{\sqrt[3]{2}}{\sqrt[3]{8}} = \dfrac{1.260}{2} = .6300$

(g) $\sqrt[3]{\dfrac{5}{9}} = \dfrac{\sqrt[3]{5}}{\sqrt[3]{9}} = \dfrac{\sqrt[3]{5}}{\sqrt[3]{9}} \cdot \dfrac{\sqrt[3]{3}}{\sqrt[3]{3}} = \dfrac{\sqrt[3]{5}\,\sqrt[3]{3}}{\sqrt[3]{27}}$

$= \dfrac{(1.710)(1.442)}{3} = \dfrac{2.465820}{3}$

$= \dfrac{2.466}{3} = .8220$

Name: _____

Class: _____

Exercises for Sec. 12-5 Estimate each root using $\sqrt{2} = 1.414$, $\sqrt{3} = 1.732$, $\sqrt{5} = 2.236$, or $\sqrt{7} = 2.646$. Round each answer to the appropriate number of significant digits.

1. $\sqrt{\dfrac{1}{4}} =$ 2. $\sqrt{\dfrac{1}{9}} =$ 3. $\sqrt{\dfrac{1}{16}} =$

4. $\sqrt{\dfrac{1}{25}} =$ 5. $\sqrt{\dfrac{9}{16}} =$ 6. $\sqrt{\dfrac{16}{25}} =$

7. $\sqrt{\dfrac{25}{49}} =$ 8. $\sqrt{\dfrac{25}{9}} =$

9. $\sqrt{\dfrac{1}{7}} =$

10. $\sqrt{\dfrac{1}{5}} =$

11. $\sqrt{\dfrac{1}{12}} =$

12. $\sqrt{\dfrac{1}{18}} =$

13. $\sqrt{\dfrac{1}{50}} =$

14. $\sqrt{\dfrac{1}{20}} =$

15. $\sqrt{\dfrac{7}{3}} =$

16. $\sqrt{\dfrac{3}{5}} =$

17. $\sqrt{\dfrac{7}{18}} =$

18. $\sqrt{\dfrac{9}{20}} =$

19. $\sqrt{\dfrac{4}{27}} =$

20. $\sqrt{\dfrac{5}{12}} =$

Estimate each root using $\sqrt[3]{2} = 1.260$, $\sqrt[3]{3} = 1.442$, $\sqrt[3]{4} = 1.587$, $\sqrt[3]{5} = 1.710$, $\sqrt[3]{6} = 1.817$, or $\sqrt[3]{7} = 1.913$. Round each root to the appropriate number of significant digits.

21. $\sqrt[3]{\dfrac{1}{27}} =$

22. $\sqrt[3]{\dfrac{1}{8}} =$

23. $\sqrt[3]{\dfrac{3}{8}} =$

24. $\sqrt[3]{\dfrac{4}{27}} =$

25. $\sqrt[3]{\dfrac{2}{27}} =$

26. $\sqrt[3]{\dfrac{7}{8}} =$

27. $\sqrt[3]{\dfrac{6}{8}} =$

28. $\sqrt[3]{\dfrac{5}{27}} =$

29. $\sqrt[3]{\dfrac{6}{27}} =$

30. $\sqrt[3]{\dfrac{4}{8}} =$

31. $\sqrt[3]{\dfrac{1}{3}} =$

32. $\sqrt[3]{\dfrac{1}{2}} =$

33. $\sqrt[3]{\dfrac{4}{9}} =$

34. $\sqrt[3]{\dfrac{8}{9}} =$

35. $\sqrt[3]{\dfrac{27}{4}} =$

36. $\sqrt[3]{\dfrac{1}{9}} =$

37. $\sqrt[3]{\dfrac{6}{9}} =$

38. $\sqrt[3]{\dfrac{3}{4}} =$

39. $\sqrt[3]{\dfrac{2}{3}} =$

40. $\sqrt[3]{\dfrac{3}{2}} =$

Unit 12 Test Simplify each product so that the base appears once with a positive exponent.

1. $(8^9)(8^{13}) =$
2. $(7^{-15})(7^{-5}) =$
3. $(5^9)(5^{-12}) =$

Simplify each product so that each base appears once with a positive exponent.

4. $(2^{10})(3^{-5})(2^{-3})(3^{-7}) =$
5. $(8^7)(2^{15})(8^{-11})(3^{-7})(2^{-4})(3^{-6}) =$

Compute each quotient.

6. $\dfrac{2^{-9}}{2^{-12}} =$
7. $\dfrac{(8^3)(3^{12})}{(8^5)(3^9)} =$
8. $\dfrac{(2^{-5})(3^4)(5^{-5})}{(3^6)(5^{-4})(2^{-7})} =$

Compute each product or quotient using scientific notation. Round each answer to the appropriate number of significant digits.

9. $63{,}000{,}000 \times 3{,}000{,}000{,}000 =$

10. $.000027 \times 8{,}100{,}000{,}000{,}000 =$

11. $\dfrac{.000058}{.00000006} =$

12. $\dfrac{490{,}000{,}000{,}000}{.0000000034} =$

Estimate each root using $\sqrt{2} = 1.414$, $\sqrt{11} = 3.317$, $\sqrt{13} = 3.606$, or $\sqrt{17} = 4.123$. Round each answer to the appropriate number of significant digits.

13. $\sqrt{44} =$

14. $\sqrt{117} =$

15. $\sqrt{\dfrac{17}{36}} =$

16. $\sqrt{\dfrac{11}{8}} =$

Estimate each root using $\sqrt[3]{2} = 1.260$, $\sqrt[3]{11} = 2.224$, $\sqrt[3]{13} = 2.351$, or $\sqrt[3]{17} = 2.571$. Round each answer to the appropriate number of significant digits.

17. $\sqrt[3]{104} =$

18. $\sqrt[3]{297} =$

19. $\sqrt[3]{\dfrac{17}{27}} =$

20. $\sqrt[3]{\dfrac{13}{4}} =$

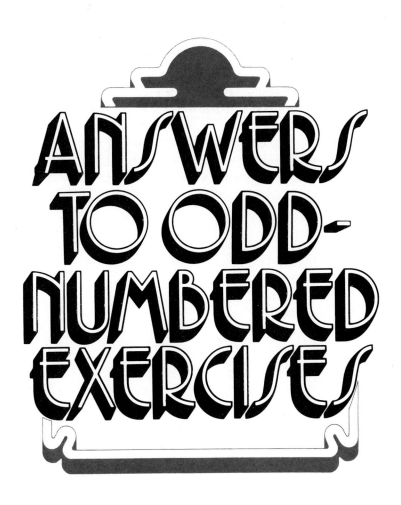

Unit 1

Section 1-1
1. $20 + 9 = 2 \times 10^1 + 9$
3. $700 + 30 + 4 = 7 \times 10^2 + 3 \times 10^1 + 4$
5. $900 + 30 = 9 \times 10^2 + 3 \times 10^1$
7. $6,000 + 80 + 1 = 6 \times 10^3 + 8 \times 10^1 + 1$
9. $5,000 + 600 + 90 + 7 = 5 \times 10^3 + 6 \times 10^2 + 9 \times 10^1 + 7$
11. $50,000 + 5,000 + 90 + 1 = 5 \times 10^4 + 5 \times 10^3 + 9 \times 10^1 + 1$
13. $90,000 + 2,000 + 7 = 9 \times 10^4 + 2 \times 10^3 + 7$
15. $300,000 + 70,000 + 5,000 + 200 + 10 + 6$
    $= 3 \times 10^5 + 7 \times 10^4 + 5 \times 10^3 + 2 \times 10^2 + 10^1 + 6$
17. $100,000 + 200 + 50 + 3 = 10^5 + 2 \times 10^2 + 5 \times 10^1 + 3$
19. $8,000,000 + 700,000 + 50,000 + 2,000 + 500 + 40 + 6$
    $= 8 \times 10^6 + 7 \times 10^5 + 5 \times 10^4 + 2 \times 10^3 + 5 \times 10^2 + 4 \times 10^1 + 6$

Section 1-2
1. 59
3. 147
5. 165
7. 578
9. 838
11. 1,651
13. 189
15. 160
17. 218
19. 1,852
21. 2,131
23. 32
25. 244
27. 145
29. 32
31. 287
33. 1,942
35. 39
37. 1,667
39. 364

Section 1-3
1. 11
3. 18
5. 25
7. 3
9. 13
11. 47
13. 324
15. 292
17. 422
19. 8
21. 5,513
23. 1,799
25. 272
27. 999
29. 33,204

Section 1-4
1. 48
3. 96
5. 342
7. 684
9. 1,728
11. 2,412
13. 4,446
15. 2,919
17. 34,580
19. 12,048
21. 58,976

Section 1-5
1. 15
3. 92
5. 206
7. $145\frac{5}{6}$
9. 23
11. 22
13. 32
15. 51
17. $50\frac{8}{67}$
19. $70\frac{53}{96}$
21. 343

Section 1-4 (continued)
23. 500,365
25. 706,742
27. 2,882,392
29. 1,015,245

Section 1-5 (continued)
23. 730
25. $852\frac{23}{47}$
27. 25
29. $53\frac{41}{873}$

Unit 1 Test
1. $90 + 6 = 9 \times 10^1 + 6$
2. $700 + 20 + 9 = 7 \times 10^2 + 2 \times 10^1 + 9$
3. $8,000 + 30 + 2 = 8 \times 10^3 + 3 \times 10^1 + 2$
4. $40,000 + 3,000 + 900 + 60 + 5 = 4 \times 10^4 + 3 \times 10^3 + 9 \times 10^2 + 6 \times 10^1 + 5$
5. 121
6. 2,074
7. 216
8. 2,531
9. 16
10. 226
11. 316
12. 60
13. 342
14. 3,816
15. 28,782
16. 132,966
17. $14\frac{3}{6}$ or $14\frac{1}{2}$
18. $76\frac{2}{7}$
19. $80\frac{5}{37}$
20. $323\frac{53}{84}$

Unit 2

Section 2-1
1. (2)(3)
3. (2)(2)(2)
5. (2)(2)(2)(2)
7. (2)(2)(5)
9. (2)(2)(2)(3)
11. (2)(2)(7)
13. (2)(2)(2)(5)
15. (2)(2)(3)(5)
17. (2)(5)(7)
19. (2)(2)(2)(2)(5)
21. (2)(2)(3)(7)
23. (2)(2)(5)(5)
25. (11)(11)
27. (2)(2)(2)(19)
29. (2)(2)(3)(3)(5)
31. (2)(2)(2)(3)(3)(3)
33. (2)(2)(2)(2)(2)(3)(3)
35. (2)(2)(2)(3)(3)(5)
37. (2)(2)(2)(2)(2)(2)(5)
39. (2)(2)(2)(2)(2)(5)(5)

Section 2-2
1. $\frac{2}{3}$
3. $\frac{3}{5}$
5. $\frac{5}{6}$
7. $\frac{3}{2}$
9. $\frac{2}{3}$
11. $\frac{3}{5}$
13. $\frac{1}{2}$
15. $\frac{3}{2}$
17. $\frac{9}{7}$
19. $\frac{7}{8}$
21. 2
23. $\frac{25}{36}$
25. $\frac{16}{19}$
27. $\frac{9}{14}$
29. $\frac{5}{6}$

Section 2-3
1. $\frac{10}{15}$
3. $\frac{18}{27}$
5. $\frac{15}{20}$
7. $\frac{15}{25}$
9. $\frac{27}{72}$
11. $\frac{12}{42}$
13. $\frac{100}{180}$
15. $\frac{35}{60}$
17. $\frac{35}{91}$
19. $\frac{9}{87}$
21. $\frac{136}{152}$
23. $\frac{27}{141}$
25. $\frac{135}{252}$
27. $\frac{124}{168}$
29. $\frac{328}{360}$
31. $\frac{102}{288}$
33. $\frac{153}{504}$
35. $\frac{297}{576}$
37. $\frac{248}{648}$
39. $\frac{115}{480}$

Section 2-4
1. 45
3. 24
5. 90
7. 48
9. 336
11. 84
13. 96
15. 336
17. 1,728
19. 216
21. 432
23. 60
25. 72
27. 144
29. 180
31. 360
33. 588
35. 64
37. 192
39. 864

Unit 2 Test
1. (2)(7)
2. (2)(2)(5)
3. (2)(2)(3)(3)
4. (3)(3)(3)(3)
5. (2)(2)(2)(3)(3)(3)
6. 24
7. 140
8. 280
9. 72
10. 360
11. $\frac{1}{3}$
12. $\frac{3}{4}$
13. $\frac{9}{5}$
14. $\frac{4}{3}$
15. $\frac{3}{4}$
16. $\frac{16}{36}$
17. $\frac{63}{72}$
18. $\frac{95}{120}$
19. $\frac{35}{270}$
20. $\frac{77}{504}$

Unit 3

Section 3-1
1. $\frac{1}{2}$
3. $\frac{6}{35}$
5. $\frac{7}{10}$

Section 3-2
1. $\frac{3}{2}$
3. $\frac{55}{27}$
5. $\frac{1}{2}$

Section 3-1 (continued)
7. $\frac{5}{8}$
9. 2
11. $\frac{1}{4}$
13. $\frac{3}{8}$
15. $\frac{1}{3}$
17. $\frac{16}{3}$
19. 1
21. $\frac{1}{3}$
23. 1
25. 4
27. $\frac{3}{8}$
29. $\frac{1}{10}$
31. $\frac{2}{3}$
33. $\frac{1}{36}$
35. $\frac{5}{18}$
37. $\frac{1}{2}$
39. $\frac{4}{9}$

Section 3-2 (continued)
7. 5
9. $\frac{2}{3}$
11. $\frac{2}{5}$
13. $\frac{2}{3}$
15. $\frac{4}{3}$
17. $\frac{5}{8}$
19. $\frac{1}{2}$
21. 16
23. $\frac{27}{20}$
25. $\frac{1}{6}$
27. 1
29. 2
31. $\frac{3}{2}$
33. $\frac{1}{3}$
35. $\frac{3}{4}$
37. $\frac{28}{27}$
39. $\frac{7}{3}$
41. $\frac{2}{3}$
43. 2
45. $\frac{1}{2}$
47. $\frac{2}{3}$

Section 3-3
1. $\frac{7}{9}$
3. 1
5. $\frac{3}{2}$
7. $\frac{2}{3}$
9. $\frac{19}{12}$
11. $\frac{13}{18}$
13. $\frac{31}{36}$
15. $\frac{17}{30}$
17. $\frac{25}{36}$
19. $\frac{53}{60}$
21. $\frac{89}{150}$
23. $\frac{11}{24}$
25. $\frac{23}{70}$
27. $\frac{71}{108}$
29. $\frac{109}{216}$
31. $\frac{9}{4}$
33. $\frac{13}{8}$
35. $\frac{119}{72}$
37. $\frac{5}{6}$
39. $\frac{19}{21}$
41. $\frac{143}{240}$
43. $\frac{97}{216}$
45. $\frac{145}{648}$
47. $\frac{27}{16}$

Section 3-4
1. $\frac{5}{7}$
3. $\frac{1}{4}$
5. $\frac{1}{2}$
7. $\frac{1}{6}$
9. $\frac{7}{18}$
11. $\frac{13}{36}$
13. $\frac{5}{36}$
15. $\frac{3}{20}$
17. $\frac{1}{24}$
19. $\frac{1}{84}$
21. $\frac{1}{3}$
23. $\frac{5}{72}$
25. $\frac{5}{54}$
27. $\frac{1}{30}$
29. $\frac{5}{216}$

Section 3-3 (continued)
49. $\frac{16}{9}$
51. $\frac{127}{72}$
53. $\frac{181}{144}$
55. $\frac{217}{432}$

Unit 3 Test
1. $\frac{10}{21}$
2. $\frac{5}{14}$
3. $\frac{1}{6}$
4. $\frac{9}{8}$
5. $\frac{3}{4}$
6. $\frac{21}{10}$
7. $\frac{8}{7}$
8. $\frac{1}{6}$
9. $\frac{50}{81}$
10. $\frac{3}{8}$
11. $\frac{5}{7}$
12. $\frac{47}{36}$
13. $\frac{31}{36}$
14. $\frac{77}{108}$
15. $\frac{49}{72}$
16. $\frac{3}{13}$
17. $\frac{4}{35}$
18. $\frac{9}{20}$
19. $\frac{5}{96}$
20. $\frac{5}{216}$

Unit 4

Section 4-1
1. $\frac{5}{2}$
3. $\frac{11}{3}$
5. $\frac{48}{7}$
7. $\frac{39}{8}$
9. $\frac{37}{10}$
11. $\frac{43}{3}$
13. $\frac{63}{4}$
15. $\frac{163}{5}$
17. $\frac{107}{2}$
19. $\frac{377}{6}$
21. $\frac{41}{12}$
23. $\frac{63}{10}$
25. $\frac{173}{24}$
27. $\frac{23}{20}$
29. $\frac{223}{54}$

Section 4-2
1. $1\frac{1}{2}$
3. $3\frac{1}{2}$
5. $1\frac{1}{4}$
7. $2\frac{2}{3}$
9. $1\frac{1}{5}$
11. $1\frac{5}{7}$
13. $5\frac{2}{3}$
15. $1\frac{4}{5}$
17. $2\frac{5}{6}$
19. $5\frac{1}{3}$
21. $1\frac{3}{5}$
23. $1\frac{12}{13}$
25. $2\frac{2}{3}$
27. $34\frac{1}{2}$
29. $9\frac{1}{7}$
31. $5\frac{3}{5}$
33. $41\frac{2}{3}$
35. $73\frac{2}{7}$
37. $4\frac{7}{13}$

Section 4-2 (continued)
39. $2\frac{25}{29}$
41. $3\frac{8}{23}$
43. $3\frac{5}{17}$
45. $3\frac{20}{51}$
47. $5\frac{72}{95}$
49. $4\frac{170}{173}$

Section 4-3
1. 12
3. $3\frac{1}{3}$
5. $6\frac{2}{3}$
7. $7\frac{1}{2}$
9. $9\frac{3}{7}$
11. $3\frac{3}{4}$
13. 27
15. $10\frac{2}{3}$
17. $17\frac{1}{2}$
19. $4\frac{4}{9}$
21. 6
23. $2\frac{1}{3}$
25. $2\frac{1}{5}$
27. $10\frac{2}{7}$
29. $7\frac{1}{2}$
31. $3\frac{3}{4}$
33. 2
35. $1\frac{2}{5}$
37. $1\frac{1}{5}$
39. $1\frac{1}{8}$
41. $1\frac{1}{15}$
43. $\frac{3}{4}$
45. $1\frac{1}{7}$
47. 9
49. $1\frac{1}{4}$
51. $1\frac{1}{14}$
53. $\frac{25}{32}$
55. $1\frac{3}{7}$

Section 4-4
1. 7
3. $13\frac{3}{4}$
5. $6\frac{5}{6}$
7. $16\frac{1}{9}$
9. $16\frac{7}{36}$
11. $31\frac{59}{72}$
13. $75\frac{77}{100}$
15. $94\frac{47}{48}$
17. $83\frac{7}{48}$
19. $116\frac{32}{147}$
21. $87\frac{13}{60}$
23. $121\frac{1}{96}$
25. $142\frac{155}{432}$
27. $11\frac{19}{36}$
29. $19\frac{7}{24}$
31. $16\frac{2}{3}$
33. $25\frac{4}{15}$
35. $12\frac{93}{100}$
37. $47\frac{191}{588}$
39. $76\frac{27}{28}$

Section 4-5
1. $1\frac{1}{2}$
3. $3\frac{4}{13}$
5. $\frac{4}{7}$
7. $2\frac{3}{4}$
9. $5\frac{17}{18}$
11. $8\frac{7}{60}$
13. $7\frac{41}{45}$
15. $49\frac{19}{60}$
17. $1\frac{25}{126}$
19. $12\frac{2}{15}$

Applications
1. 192 mi
3. 5 trips
5. 4,100 ft²
7. $\frac{5}{8}$ lb
9. $9\frac{9}{14}$ min
11. $3\frac{3}{4}$ hr
13. $4
15. $25\frac{3}{8}$ gal
17. $\frac{3}{500}$
19. $\frac{1}{3}$ lb

Section 4-5 (continued)

21. $29\frac{7}{120}$
23. $15\frac{5}{144}$
25. $62\frac{1}{10}$
27. $27\frac{17}{108}$
29. $14\frac{377}{392}$

Applications (continued)

21. $2\frac{9}{20}$ hr
23. $\frac{11}{64}$ in
25. $\frac{4}{5}$
27. $3\frac{2}{15}$ gal/min
29. $\frac{4}{15}$
31. $\frac{2}{9}$
33. 241¢
35. $\frac{11}{32}$ in

Unit 4 Test

1. $\frac{68}{9}$
2. $\frac{107}{8}$
3. $2\frac{2}{3}$
4. $14\frac{2}{9}$
5. $13\frac{1}{2}$
6. 28
7. 14
8. $\frac{1}{2}$
9. $1\frac{1}{3}$
10. $1\frac{5}{9}$
11. $13\frac{1}{2}$
12. $9\frac{1}{6}$
13. $7\frac{43}{72}$
14. $17\frac{43}{72}$
15. $2\frac{1}{3}$
16. $\frac{5}{8}$
17. $6\frac{61}{72}$
18. 20 gal
19. $4
20. $16\frac{7}{12}$ lb

Unit 5

Section 5-1

1. $\dfrac{23}{100} = \dfrac{2}{10} + \dfrac{3}{100}$

3. $\dfrac{578}{100} = 5 + \dfrac{7}{10} + \dfrac{8}{100}$

5. $\dfrac{46{,}941}{1{,}000} = 46 + \dfrac{9}{10} + \dfrac{4}{100} + \dfrac{1}{1{,}000}$

7. $\dfrac{73{,}003}{1{,}000} = 73 + \dfrac{3}{1{,}000}$

9. $\dfrac{70{,}003}{100} = 700 + \dfrac{3}{100}$

11. $.17 = \dfrac{1}{10} + \dfrac{7}{100}$

13. $.00632 = \dfrac{6}{1,000} + \dfrac{3}{10,000} + \dfrac{2}{100,000}$

15. $7.05 = 7 + \dfrac{5}{100}$

17. $82.895 = 80 + 2 + \dfrac{8}{10} + \dfrac{9}{100} + \dfrac{5}{1,000}$

19. $602.03 = 600 + 2 + \dfrac{3}{100}$

Section 5-2
1. 1.92
3. .0247
5. 21.9
7. 18.16
9. 20.731
11. 117.16
13. 1,474.9
15. .2011
17. 8.839
19. 25.819
21. .7
23. 58.6
25. .16
27. 3.53
29. .09
31. 1.86
33. 3.168
35. .09
37. 6.65
39. 9.99

Section 5-3
1. 8.96
3. .896
5. .000896
7. .05664
9. .1264
11. 17.43
13. .03680
15. .004543
17. 2.3296
19. 24.586
21. 540.72
23. 298.41
25. 3.6754
27. .09408
29. .0017986

Section 5-4
1. 8.479   8.48   8.5
3. 37.982 37.98 38.0
5. 4.827   4.83   4.8
7. .631   .63   .6
9. .000   .00   .0
11. 62.701 62.70 62.7
13. .403   .40   .4
15. 37.943 37.94 37.9
17. 5.9275
19. 1.0000
21. 376   380   400
23. 94   100   100
25. 983   980   1,000
27. 6,750 6,750 6,800
29. 9,000 9,000 9,000

Answers to Odd-Numbered Exercises

Section 5-5
1. .87
3. 8.7
5. .087
7. .258
9. .820
11. .05
13. .000
15. 1,256.667
17. .24
19. .003
21. 420
23. .87
25. 288.889
27. .001
29. 41.933

Section 5-6
1. .5
3. .6
5. .667
7. .444
9. .125
11. 1.125
13. .167
15. 2.571
17. .417
19. .35
21. 2.389
23. .138
25. 2.969
27. .028
29. .002
31. 5.75
33. 2.667
35. 4.625
37. 3.385
39. 7.208

Applications
1. 468.085 mi per hr
3. $255.13
5. $325.44
7. 1.69 times as heavy
9. 25.2 gal per min
11. $1.39
13. 6.5 hr
15. 57.803 sec
17. 450 shares
19. 6,749.256 lb
21. 90.929 yd per game
23. $343.75
25. .381 in
27. .59375 in
29. 150 times

Unit 5 Test
1. $\frac{56}{1,000} = \frac{5}{100} + \frac{6}{1000}$
2. $\frac{7,382}{100} = 70 + 3 + \frac{8}{10} + \frac{2}{100}$
3. $42.53 = 42\frac{53}{100}$
4. 10.0
5. 13.6
6. 45.006
7. 3.18
8. 488.64
9. .86
10. 3.0
11. 23.97
12. .510
13. 3.484

Answers to Odd-Numbered Exercises

14. 6
15. 5.3
16. .21
17. .032
18. 7.4 tons
19. 56.2 mi per hr
20. $6.50

Unit 6

Section 6-1
1. 4
3. 8
5. 11
7. $2\frac{1}{4}$
9. $5\frac{1}{15}$
11. $2\frac{1}{4}$
13. $\frac{3}{20}$
15. $1\frac{1}{2}$
17. 12
19. $1\frac{1}{5}$
21. $3\frac{1}{3}$
23. $7\frac{1}{2}$
25. 16
27. 1
29. 2
31. $\frac{2}{3}$
33. $\frac{5}{8}$
35. $\frac{2}{3}$
37. $\frac{2}{3}$
39. $2\frac{1}{2}$

Section 6-2
1. 2
3. 2
5. 12
7. 15
9. 30
11. 0
13. 3
15. 8.8
17. 4.2
19. $\frac{2}{7}$
21. $1\frac{1}{3}$
23. $\frac{1}{4}$
25. $\frac{13}{15}$
27. 1.1
29. $\frac{7}{36}$

Unit 6 Test
1. 3
2. $2\frac{1}{3}$
3. $1\frac{1}{3}$
4. 4
5. 15
6. 20
7. $2\frac{1}{2}$
8. $\frac{3}{4}$
9. $7\frac{1}{2}$
10. 14
11. $1\frac{7}{33}$
12. $2\frac{1}{3}$
13. 13
14. 14
15. 1.4

16. 7.2
17. $\frac{1}{3}$
18. $\frac{2}{3}$
19. $\frac{5}{12}$
20. $\frac{7}{18}$

Unit 7

Section 7-1

1. .07
3. .39
5. .052
7. .837
9. .006
11. 1.85
13. 4.563
15. .075
17. .0575
19. .0867
21. $\frac{3}{400}$
23. $\frac{7}{300}$
25. $\frac{27}{500}$
27. $\frac{69}{800}$
29. $\frac{51}{400}$
31. $\frac{251}{200}$
33. $\frac{1}{20}$
35. $\frac{173}{100}$
37. $\frac{21}{500}$
39. $\frac{9}{1,000}$
41. 50%
43. 72%
45. 6%
47. 52.9%
49. 7.8%
51. .75%
53. 40%
55. 37.5%
57. 33.3%
59. 28.6%
61. 50%
63. $37\frac{1}{2}$%
65. $33\frac{1}{3}$%
67. $2\frac{1}{2}$%
69. $\frac{1}{2}$%
71. $\frac{2}{5}$%
73. $\frac{1}{5}$%
75. $2\frac{1}{2}$%
77. $25\frac{3}{10}$%
79. $35\frac{47}{100}$%

Section 7-2

1. 4.5
3. 75%
5. 900
7. 25
9. 5.2
11. 4.8
13. $\frac{2}{5}$
15. 3%
17. 37.5%
19. 60
21. $\frac{2}{3}$%
23. 200
25. 100
27. 60
29. 60

Section 7-3

1. 7.15 lb
3. 16.35%
5. (a) $41\frac{2}{3}$% or 41.7%
   (b) $58\frac{1}{3}$% or 58.3%
7. 62%
9. 42.9%
11. 40%
13. 128 lb
15. 23 bearings
17. 927 seeds
19. 300 calories
21. $1.11
23. $164.96
25. $73,000,000
27. 55%

Section 7-2 (continued)
31. 5
33. 750
35. 2.5
37. 160%
39. 20

Section 7-3 (continued)
29. 32.4%
31. $64.00
33. $1,200,000
35. $2.70
37. $40,000,000
39. (a) $12,500
    (b) 125
    (c) $502.50

Unit 7 Test
1. .27
2. .084
3. .054
4. $\dfrac{1}{30}$
5. $\dfrac{23}{1,000}$
6. 2.3%
7. 55.6%
8. .4%
9. $24\dfrac{3}{10}\%$
10. $87\dfrac{1}{2}\%$
11. $\dfrac{1}{5}\%$
12. 40
13. 12.5%
14. 32.4
15. 50
16. 37.5%
17. 96
18. 6%
19. 5,000 lb
20. 120 lb

Unit 8

Section 8-1
1. $5 > 3$
3. $0 < 2$
5. $-5 < 2$
7. $3 > -4$
9. $-6 > -9$
11. $-8 < -2$
13. $\tfrac{1}{3} < \tfrac{1}{2}$
15. $-\tfrac{3}{4} < -\tfrac{1}{2}$
17. $3.2 < 3.25$

19. $4.6 > -5.9$
21. $1 < 5 < 7 < 12$
23. $-6 < -2 < 0 < 5 < 12$
25. $-7 < -5 < -4 < -1 < 1 < 3 < 7$
27. $-.9 < -.82 < -.21 < -.05 < .19 < .3 < .31$
29. $-\frac{3}{4} < -\frac{1}{5} < -\frac{1}{6} < \frac{1}{10} < \frac{1}{4} < \frac{5}{7}$
31. $7 > 5 > 4 > 1$
33. $7 > 2 > 0 > -3 > -5$
35. $7 > 2 > 1 > -1 > -3 > -4 > -5$
37. $9 > 4 > 3 > 1 > 0 > -1 > -2 > -4 > -5$
39. $\frac{5}{6} > \frac{1}{3} > \frac{2}{15} > -\frac{1}{2} > -\frac{5}{8} > -\frac{8}{9}$

Section 8-2
1. 2
3. 2
5. 3
7. 3
9. -15
11. -5
13. 0
15. -5
17. -1
19. 0
21. 5
23. 1
25. -8
27. -6
29. 47

Section 8-3
1. $-2 + 5 + (-6)$
3. $6 + (-5) + 7 + (-8)$
5. $4 + (-3) + (-4)$
7. $-5 + (-2) + (-3) + 1$
9. $-3 + 5 + (-8) + (-4) + (-6)$
11. -4
13. -6
15. -34
17. -4
19. -9
21. -8
23. 4
25. -6
27. -3
29. -20
31. 0
33. 11
35. -27
37. -20
39. -4

Section 8-4
1. 15
3. -12
5. -12
7. -48
9. 72
11. -24
13. -30
15. 36
17. 60
19. -72
21. 120
23. -72
25. -2
27. 3
29. -6
31. $-1\frac{1}{2}$
33. $1\frac{1}{4}$
35. $-1\frac{1}{2}$
37. $\frac{2}{3}$
39. $-\frac{4}{5}$
41. $\frac{1}{3}$
43. -9
45. -1
47. $2\frac{2}{3}$
49. $-1\frac{1}{8}$

Unit 8 Test
1. $-2 < 5$
2. $-2 > -7$
3. $-6 < -3 < -2 < 0 < 1 < 4 < 7$

4. 2
5. −1
6. −7
7. −4
8. 13
9. −9
10. −5
11. 54
12. −40
13. −24
14. −5
15. −4
16. $1\frac{1}{2}$
17. $1\frac{1}{2}$
18. $-\frac{4}{5}$
19. −2
20. 3

Unit 9

Section 9-1
1. 12
3. −18
5. −8
7. 11
9. −21
11. 1
13. 9
15. 0
17. 16
19. −21
21. −18
23. −12
25. −25
27. 13
29. 26
31. 31
33. −18
35. −4
37. −2
39. −16

Section 9-2
1. 36
3. 6
5. 35
7. 21
9. 1
11. −1
13. 1
15. 40
17. 2
19. 7
21. 91
23. −32
25. 6
27. 41
29. 22

Section 9-3
1. −3
3. $\frac{1}{2}$
5. $\frac{3}{4}$
7. −2
9. $-2\frac{1}{2}$
11. $-1\frac{1}{6}$
13. $\frac{5}{18}$
15. $-\frac{8}{9}$
17. $\frac{1}{16}$
19. $\frac{1}{24}$
21. −3
23. $\frac{1}{12}$
25. $5\frac{3}{5}$
27. $1\frac{1}{2}$
29. −1

Unit 9 Test
1. −4
2. 4
3. 2
4. 2
5. 8
6. $9\frac{1}{3}$
7. $-25\frac{1}{2}$

8. 2
9. 50
10. 8
11. 63
12. −18
13. 12
14. −22
15. 8
16. −2
17. $\frac{2}{3}$
18. $-\frac{1}{2}$
19. $\frac{1}{3}$
20. $2\frac{1}{5}$

Unit 10

Section 10-1
1. 24
3. 14
5. 22
7. 21
9. 15
11. 16
13. 62
15. 9
17. 125
19. 24
21. $2\frac{2}{3}$
23. $1\frac{4}{5}$
25. 18
27. 216
29. 72
31. $\frac{3}{25}$
33. $-3\frac{1}{2}$
35. 3
37. $3\frac{1}{3}$
39. $13\frac{1}{2}$

Section 10-2
1. 10
3. 34
5. 36
7. 2
9. 8
11. −6
13. $1\frac{5}{18}$
15. 0
17. $\frac{4}{15}$
19. 80
21. 24
23. 44
25. $2\frac{1}{2}$
27. 34
29. $-\frac{1}{64}$

Section 10-3
1. $A + 5$
3. $A + 5$
5. $A - 5$
7. $A + B$
9. AB
11. $\dfrac{A}{B}$
13. $C - A$
15. $A^2 B$
17. $A^2 - B$

19. $\left(\dfrac{A}{B}\right)^2$

21. $(A - B)^3$

23. $A(B - C)$

25. $\dfrac{LK}{J}$

27. $\dfrac{L}{KJ}$

29. $\dfrac{K}{J} + L$

31. $(A - B)^3 - 5$

33. $A^2 + B + 5$

35. $R - \dfrac{SW}{T}$

37. $\dfrac{AB - 6}{A - B}$

39. $\dfrac{A + BC}{RST}$

Section 10-4

1. $R = P + Q$
3. $Q = 2(A - B)$
5. $M = A(3B + C)$
7. $J = MN + 6$
9. $S = M + N - 5$
11. $D = \dfrac{1}{2}(A + C^2)$
13. $N = \dfrac{A + B}{A - B}$
15. $V = \dfrac{4}{3}\pi R^3$
17. $d = \dfrac{W}{V}$
19. $E = IR$
21. $i = prt$
23. $P = 2l + 2w$
25. Mary's age = 2 × Alice's age
27. Carol's age = Susan's age + Anne's age
29. Cost = selling price − margin
31. Selling price = list price − markdown
33. Sales tax = 6% × selling price
35. Distance = average speed × hours driven
37. Price per pound = $\dfrac{\text{price}}{\text{number of pounds}}$
39. Wage = hourly rate × hours worked

Section 10-5
1. 3 years
3. 6 years
5. 215 lb
7. $325
9. $210
11. 200 acres
13. 15 mi/gal
15. $240
17. $46.67
19. 3,150 watts
21. 22 ohms
23. .313
25. 3.50

Unit 10 Test
1. 17
2. 20
3. −6
4. $4\frac{1}{6}$
5. 24
6. −2
7. $2\frac{1}{2}$
8. 36
9. $2(a + b)$
10. $mn - 5$
11. $(r - s)^3$
12. $\dfrac{cd}{c + d}$
13. $K = a + b$
14. $L = mn^2$
15. $M = \dfrac{r + s}{t}$
16. $V = a(a + b)^2$
17. Bob's age = 2 × Jim's age + 4
18. Running average = $\dfrac{\text{yards gained} - \text{yards lost}}{\text{times ball carried}}$
19. 10 years
20. $60

Unit 11

Section 11-1
1. 120¢
3. $12
5. 80 oz
7. $.63
9. $160
11. 40¢
13. $2\frac{1}{4}$ lb

Section 11-2
1. 5.9 mm
3. 12.75 m
5. 4.379 km
7. 734 mm
9. 5.274 km
11. 53 mg
13. 5.325 kg

Answers to Odd-Numbered Exercises

Section 11-1 (continued)
15. 60 in
17. $.99
19. $7\frac{1}{2}$ oz
21. 81 ft
23. $50\frac{2}{5}$ gal
25. $9\frac{3}{5}$ hr
27. 40 runs
29. $24
31. (a) 45
    (b) 28%
    (c) 225
33. $240
35. $1\frac{1}{9}$ ¢
37. $180
39. (a) 2,100 lb
    (b) 1,944 lb

Section 11-2 (continued)
15. .458 g
17. 6,512 mg
19. 57.309128 kg
21. 7,280 cm²
23. 5.852931 m²
25. 82.342890 km²
27. 60,000 cm²
29. .2 cm²
31. 5,000,000,000 m³
33. 5 cm³
35. 30,000,000 cm³
37. .005 liter
39. 6 m³
41. 8 m 26 cm 7 mm
43. 29 kg 168 g 803 mg
45. 14 km 546 m 52 mm
47. 6 kg 500 g 964 mg
49. 89 cm 61 mm

Section 11-3
1. 10 in
3. $1\frac{1}{3}$ mi
5. $31\frac{1}{2}$ in
7. $29\frac{1}{3}$ in
9. $2\frac{1}{4}$ lb
11. $4\frac{2}{3}$ oz
13. 1,500 lb
15. $\frac{3}{5}$ ton
17. 5 ft²
19. 24 in²
21. 27,878,400 ft²
23. 54 ft²
25. $2\frac{1}{4}$ ft³
27. $\frac{5}{16}$ yd³
29. $\frac{2}{3}$ pt
31. 9.6 ft³
33. 12 gal 3 qt 1 pt
35. 20 yd 3 in
37. 1 lb 11 oz
39. 2 yd 2 ft 10 in

Section 11-4
1. .79 in
3. 127 mm
5. .635 m
7. 39.4 ft
9. 160.9 km
11. 13.2 lb
13. 10,000 g
15. .34 kg
17. .176 oz
19. 28,409 mg
21. 12.9 cm²
23. 1.6 in²
25. 1.29 m²
27. 928.8 cm²
29. .84 m²
31. 6.10 in³
33. 28,321.92 cm³
35. 1.2 in³
37. .9 liter
39. 1.1 ft³

Section 11-5
1. .06 m/min
5. 96.6 km/hr
9. 1,288 km
13. 8.3 min
17. 1,000 kg/m³
21. 1.7 g/cm³
3. $29\frac{1}{3}$ yd/sec
7. .5 m/sec
11. 5.21 mi
15. 12,150 lb/yd³
19. 167 lb/yd³
23. 14,580 lb

25. 647 lb
27. 432 in³
29. 15 mg/cm³
31. $\frac{1}{300}$ grain/m℧
33. .25 cm³
    or $\frac{1}{320}$ grain/m℧
35. .25 cm³
37. $1.08/yd
39. $16,100,000/mi
41. $.18/ft²
43. $500/m³
45. $.80/lb
47. $22/kg
49. $5.56
51. $61,290
53. $24,800
55. $350.90

Unit 11 Test (Sections 1 through 4)

1. 7.5 kg
2. 37,500 cm
3. .001 m³
4. 3.572 liter
5. 30 m³
6. 12 oz
7. $\frac{1}{8}$ mi
8. 7.5 gal
9. 9 ft²
10. 91.4 m
11. 550 lb
12. 3.8 liter
13. 625 g
14. 12 yd 6 in
15. 22 kg 980 g 473 mg
16. 2 m 60 cm 9 mm
17. 1 liter 750 cm³
18. 20¢
19. 2,500 lb
20. 1,000¢ or $10

Unit 11 Test (Section 5)

1. 4,400 yd/min
2. 16,666.7 cm/sec
3. 5.08 cm/sec
4. 135 mi
5. 10 hr
6. 432 lb/ft³
7. .04 lb/in³
8. .36 lb/in³
9. 107.5 g
10. 5 in³
11. 5 mg/cm³
12. $\frac{1}{12}$ grain/cm³
13. $\frac{1}{3}$ mg/m℧
14. $\frac{1}{2}$ cm³ or .5 cm³
15. 10 m℧

16. $31\frac{1}{4}$ ¢/oz
17. $\frac{1}{4}$ ¢/cm³
18. 2¢/cm
19. $6,400
20. 9.7 in³

Unit 12

Section 12-1
1. 5^{10}
3. 8^{56}
5. 4^{10}
7. $(2^8)(3^9)$
9. $(2^{19})(7^8)$
11. $(2^{21})(3^8)(5^3)$
13. 25
15. $\frac{1}{216}$
17. 81
19. 12
21. $\frac{1}{63}$
23. $3\frac{1}{8}$
25. $\frac{1}{4}$
27. $\frac{3}{8}$
29. $\frac{12}{25}$

Section 12-2
1. $\dfrac{1}{3^8}$
3. 4^4
5. $\dfrac{1}{5^{13}}$
7. $\dfrac{1}{8^5}$
9. $\dfrac{1}{2}$
11. $\dfrac{1}{5^5}$
13. 10^{17}
15. $\dfrac{1}{2^3}$
17. 7^6
19. $1\dfrac{11}{16}$
21. $\dfrac{49}{125}$
23. $\dfrac{4}{27}$

25. $\dfrac{1}{16}$
27. 400
29. $5\dfrac{22}{25}$

Section 12-3
1. 3.7×10^{12}
3. 4×10^{13}
5. 3×10^3
7. 3×10^{-9}
9. 7.5×10^{-12}
11. 2.9×10^{-3}
13. 4.9×10^{-10}
15. 1.4×10^{10}
17. 1×10^{-1}
19. 5.1×10^1
21. 3.1×10^{15} mi
23. 6×10^3 times as far
25. 8×10^{11} times as great
27. 9×10^9 tons
29. 8×10^{18} kg
31. $.3 \times 10^{12}$
33. 1.8×10^3 times as heavy
35. 5.99×10^{23} atoms

Section 12-4
1. 2.828
3. 6.708
5. 5.656
7. 8.484
9. 5.292
11. 7.070
13. 9.898
15. 12
17. 11.18
19. 16.97
21. 2.520
23. 5.130
25. 3.174
27. 5.739
29. 3.826
31. 5.768
33. 6.348
35. 7.560
37. 9
39. 11.34

Section 12-5
1. $\tfrac{1}{2}$
3. $\tfrac{1}{4}$
5. $\tfrac{3}{4}$
7. $\tfrac{5}{7}$
9. .3780
11. .2887
13. .1414
15. 1.528
17. .6235
19. .3849
21. $\tfrac{1}{3}$
23. .7210
25. .4200
27. .9085
29. .6057
31. .6930
33. .7627
35. 1.890
37. .8733
39. .8733

Unit 12 Test
1. 8^{22}
2. $\dfrac{1}{7^{20}}$
3. $\dfrac{1}{5^3}$
4. $\dfrac{2^7}{3^{12}}$
5. $\dfrac{2^{11}}{(3^{13})(8^4)}$
6. 8
7. $\dfrac{27}{64}$
8. $\dfrac{4}{45}$
9. 2×10^{17}
10. 2.2×10^8
11. 1×10^3
12. 1.4×10^{20}
13. 6.634
14. 10.82
15. .6872
16. 1.173
17. 4.702
18. 6.672
19. .8570
20. 1.481

INDEX

Accuracy, 459
Area of a rectangle, 378
Associative law:
  of addition, 10
  of multiplication, 23

Base, 4, 312, 443
Base (in percent), 225
Base-10, 5

Commutative law:
  of addition, 11
  of multiplication, 23
Completeness property, 478
Concentration, 419
Conversion:
  factor of, 365
  tables, 408
Cross product, 51

Denominator, 49
  common, 89
  least common, 90
  rationalizing of, 482
Density, 418
Digit, 3
  place value of, 3, 156
Distributive law, 21
Divisibility, 27
  rules of, 41

Exponent, 4, 312, 443
Exponential form, 4, 312, 443

Factor, 3, 21, 39
  prime, 39
Factorization, 39
  prime, 39
Formula, 311
  symbolic, 341
  verbal, 341
Fraction:
  common, 49
  complex, 81
  denominator of, 49
  equivalent, 50
  improper, 50
  numerator of, 49
  proper, 50
  reciprocal of, 81
  reduced, 50
    to lowest terms, 51

Inequality, 250
Inverse additive, 347
Inverse operaton, 15

Multiple, 27
  common, 61
  least common, 61

Number:
  composite, 39
  decimal, 155
    repeating, 186
  denominate, 365
  expanded form of, 3
  factored, 39
  irrational, 186
  mixed, 115
  negative, 249
  positive, 249
  prime, 39
  rational, 186

Percent, 217, 225
Percentage, 225
Perfect cube, 475
Perfect square, 473
Power, 4, 312, 443
Precision, 460
Proportion, direct, 366

Rate, 366
Root:
  cube, 475
  square, 473

Scientific notation, 461
Significant digit, 459

Term, 281

Volume of a rectangular prism, 380